Product innovation in the Dutch food and beverage industry

Product innovation in the Dutch food and beverage industry

A study on the impact of the innovation process, strategy and network on the product's short- and long-term market performance

Christien Enzing

Innovation and sustainability series – Volume 5

Wageningen Academic
P u b l i s h e r s

Wageningen Academic Publishers,
P.O. Box 220, NL-6700 AE Wageningen,
The Netherlands.
www.WageningenAcademic.com

ISBN: 978-90-8686-131-6
ISSN: 1875-0702

First published, 2009

© Wageningen Academic Publishers
The Netherlands, 2009

Table of contents

Acknowledgements

When I started to work on this research project I didn't expect to enjoy it so much, although the subject had interested me already since I was a chemistry student and read the book 'Voedsel in Nederland' (Reijnders and Sijmons, 1974). This interest in innovation in the food and beverage industry also grew and developed from what I learned from the research I conducted at TNO in the field of technology foresight and assessment, especially in the field of life sciences and biotechnology. So when I decided to complete my academic career with a dissertation in the social sciences, the choice to focus on product innovation in the Dutch food and beverage industry was an easy one.

Working on the thesis was a period of reading, writing and discussing progress with my advisors where nobody but me dictated the pace (alongside a hectic family and work life) and where I - educated as a natural scientist - have learned a lot about the art of performing social science. Although the finalisation of the project was an effort I had to make all by myself, during the research project I had help from so many people. I would like to take the opportunity to thank them here.

My special thanks go to Onno Omta and Felix Janszen. Onno, I thank you for guiding me through this challenging academic test and stimulating me continuously to improve my work. It was a great support to me that you kept faith in the successful finalisation of the project. Felix, I am grateful for your advice at the start of the research project to choose a subject for my dissertation that would sustain my interest for a long period of time. Your practical experience with innovation management helped me interpreting the results.

In addition, I would like to thank the members of my dissertation committee: i.e. Adrie Beulens, Jan van den Ende, Alfred Kleinknecht and Ruud Smits for evaluating this thesis and participating in the defence.

I am very grateful to Sander Kern, former colleague at the TNO Institute for Strategy, Technology and Policy, who helped me in the period 1998-2000 with screening the professional and trade journals, the composition of the questionnaire and feeding the data into a database, the latter together with Jarno Dakhorst.

Special thanks go to Addy de Miranda who collected the data though the telephone interviews with the companies in 2000. I was very sorry to hear that she had passed away. I also greatly acknowledge the managers who have been interviewed and whose information is so valuable for this project. Thanks also to Sietse Sterrenburg who helped me collect the data on long-term performance and to Quirine van Wijngaarden who did a double check of the data and the references.

Furthermore, I would like to thank Maarten Batterink, Derk-Jan Haverkamp and Ron Kemp of the Business Administration Group in Wageningen for their help in processing and interpreting the statistical analyses. Talking with them and with several of their colleagues about issues and problems which I faced during the last few years of my research was very helpful.

Many people have contributed directly or indirectly to the completion of the thesis. First of all, I want to thank all my former colleagues at the TNO Innovation Policy Group for their interest and support. Special words of thanks go to Annelieke van der Giesen with whom it was very nice working together but also enjoying business trips (remember the exhibition on Impressionists in the Grand Palais?), Govert Gijsbers who was my roommate for several years and who succeeded in finalising his thesis well before me (but we still finished it in the same year!); Frans van der Zee who's decision to go to 'the other side of the road' was not without consequences (I hope your new job brings you what you missed, but I miss our discussions on books, theatre plays, concerts), Maurits Butter was a very nice colleague whose friendship I appreciated during lunches, drinks and institutes outings and last but not least Jos Leijten for insisting that I should finish the thesis sooner or later.

I thank my new colleagues Patries Boekholt and Geert van der Veen for giving me the opportunity to join the Technopolis Group; I enjoy working and lunching with them and with the other 'Technopolitans'.

This research would not have been possible without the financial support of the Agro-chain Knowledge Foundation (Agro Keten Kennis – AKK), the Dutch Organisation for Applied Technological Research (TNO) and the Dutch Ministry of Agriculture, Nature and Food Quality (LNV). This is also the place where I thank Katja Gruijters Food Design for her permission to use for the front page of this book a photo of one of the food products she has designed (photography: Jonas de Witte).

Most important was the support of Jan Willem van Wijngaarden, my husband, for whom the last eleven years have been the longest. I am very grateful that you have given me the space to pursue my interest and write the thesis. With your support at difficult moments, it was possible for me to finalise the project. Our alliance has withstood the 'thesis storm'; will the 'Alliance' cross the sea? My son Anne and my daughter Quirine have both grown up since I started working on the thesis. I am very happy that they are my 'paranimfs' during the defence. Their, and Jan Willem's love and friendship was what I enjoyed most during the long period this research project lasted.

1. Introduction

In recent years, innovation has become essential for the competitive advantage of companies. Competitiveness studies show that investments in the generation and mobilisation of new knowledge and technological skills for stimulating innovation-based growth clearly leads to better economic performance (Porter, 1985; Souder and Shermann, 1994; Christensen *et al.*, 1998; Commission, 2008). Through product and process innovation, companies are able to deliver better, faster and cheaper high-quality products and services (Tidd *et al.*, 2005). For this reason the management of innovation has become one of the core activities of innovating companies (Leonard-Barton, 1995; Burgelmann *et al.*, 2009).

Up to now, research interest in the innovation behaviour of companies has mainly focused on high-tech industries. However, innovators in low- and medium-tech industries are of comparatively greater importance for their national economies: Sandven *et al.* (2005) showed that they generate new products and processes that contribute more to the economic growth of the OECD countries than the high-tech industries. Studies on high-tech industries mostly focus on the research and development (R&D) activities as the determinant of innovation (Hirsch-Kreinsen *et al.*, 2006), while in low- and medium-tech industries many activities that lead to innovations are not R&D-based (Santamaría *et al.*, 2009). R&D activities represent about one quarter of the total expenditure on product innovations (Kleinknecht *et al.*, 2002). Robertson *et al.* (2009) conclude that there is a clear lack of understanding of the specific characteristics of innovation processes in low- and medium-tech industries. The present study tries to fill this gap by focusing on a specific low- to medium-tech industry: the food and beverage industry (F&B).

In recent years, this industry has been confronted with a number of challenges: the increased competition especially from retailers' own label products, stricter government regulations concerning food safety and recently also the economic crises leading to less spending by the consumer. These developments have created new challenges for the management of F&B companies; they have to become more pro-active and take more commercial initiatives (Martinez and Briz, 2000; Costa and Jongen, 2006). One of the best strategies for F&B companies is to be innovative. Studies confirm that companies in the F&B industry have been successful in this respect, both large firms (Robertson and Patel, 2007) and SMEs (EIM, 2008).

F&B companies are increasingly confronted with important strategic and operational questions, such as which innovation strategy will be more successful: investing in the development of new products, especially those that have proprietary elements so entry barriers can be raised, or diversifying on the basis of our core products that already have a strong market position; or a combination of both? And – as the F&B companies' resources are limited – which are the best partners with which to co-innovate, e.g. suppliers, customers or research institutions? How extensively should they be involved in the innovation process considering the potential danger of leakage of confidential information? And how can the innovation process best be organised so that all relevant resources are combined successfully? This book tries to answer these questions by discussing the important topic of product innovation in the F&B industry.

The main research question to be answered in this book is:

> *What key factors are positively related to the short- and long-term market performance of F&B products?*

In order to answer this research question a product-oriented empirical study has been carried out using data on 129 F&B products that had been announced in professional and trade journals for the Dutch F&B industry in 1998. A first wave of data collection took place mid-2000; in December 2005 data collection was completed with the data on the long-term market performance of the products. In this chapter we introduce the main research questions of the four different parts of the study presented in this book and the hypotheses that have guided the research (Section 1.4). Before that we first introduce the main theoretical concepts of the study in Section 1.1. In Section 1.2 we present the Dutch F&B industry in more detail. Section 1.3 presents the methods and the data of the study.

1.1 Innovation: process, strategy and network

1.1.1 The innovation process

Innovation is a complex phenomenon, involving the production, diffusion and translation of knowledge into new or modified products and services, as well as new production or processing techniques. We use Freeman's (1982) definition of industrial innovation that states that it is the technical design, manufacturing, management and commercial activities involved in the marketing of a new (or improved) product or the first commercial use of a new (or improved) process or equipment. Here innovation refers to the innovation process that involves the development, production and introduction of products to the market that are new or substantially improved. However, innovation is also used for describing the new item itself. In this book we focus on 'product innovations'. Product innovations may include improvements in functional characteristics or ease of use of the product. There are several types of product innovation: varying from minor modifications of existing products to products that are completely new to the firm, to the market or even to the world. Therefore, product innovations (as a new item) are often positioned according to their level of innovativeness from incremental (or evolutionary) to radical (or revolutionary) innovations. Other typologies used are sustaining against disruptive innovations (Christensen, 1997) or modular versus architectural innovations (Henderson and Clark, 1990). Besides product innovations there are two other types of innovations: process innovations and organisational innovations. The first deals with new or improved (production) processes and the latter with new or altered business structures, routines and even business models.

In the innovation process agents act to transform knowledge – through new combinations – into economic value. This can be new or existing scientific and technological knowledge, knowledge of (new and/or changing) markets, of the customers, of organisational processes and ultimately of human behaviour (Jacobs, 1999). The innovation process is considered as a dynamic, interactive and cumulative phenomenon in which actors participate from inside and outside the company. This view on the innovation process marks a significant shift from the traditional linear sequential models which prevailed until the 1980s.

Models of the innovation process

Rothwell (1994) suggests that our view on the nature of the innovation process has evolved from a simple first generation model into a complex fifth generation model. The first generation and second generation models are characterised by the dominant role of science and technology push and market pull as dominant drivers of innovation processes. In the first generation model there is a linear transfer of results of scientific and technological know-how into industrial companies that use it for the development of new products or processes (science and technology push); in the second generation model product development starts on the basis of specific market needs. Critics of these models emphasised the uncertainty about future developments as well as set-backs; they missed the feedback mechanisms between the different phases of the innovation process. Studies also showed that innovations are successful when technological expertise is combined in close relation with users (Von Hippel, 1988; Lundvall, 1992; Fagerberg, 1995). The third generation model (parallel dynamic interaction with upstream suppliers and downstream users) is very much based on the classic chain-link model introduced by Kline and Rosenberg (1986). In this model, the innovation process proceeds through small incremental steps; in each step, actors from inside and outside the company exchange information on the basis of demand and supply. After the fourth generation model (emphasising linkages and alliances), the fifth generation model (systems integration and extensive networking) is most appropriate to describe the innovation process. In this approach the different resources and actors are integrated into a coherent unity constituting the innovation system (Lundvall, 1988, 1992; Nelson, 1993). Success in innovation depends on the firm's ability to communicate and interact with a variety of external sources of knowledge (e.g. other firms, suppliers, customers, research institutions, investment companies, government agencies), as well as on the ability to co-ordinate a variety of interdependent sources of knowledge within the company (Edquist, 1997; Freeman and Soete, 1997).

However, these rather general and abstract models do not capture the developmental aspects of different industries with their specific characteristics concerning rate, type and direction of innovation. Also they differ in what enables their innovation process: this can be improving productivity facilitated by new technologies, or optimalisation of the supply chain, or specific developments on the demand side (Montalvo *et al.*, 2007). Breschi and Malerba (1997) introduced the concept of the sectoral innovation system where the boundaries of the system are endogenous, emerging from the specific context of the industry. Different industries operate under different technological regimes which are characterised by specific combinations of opportunity and appropriability conditions, degrees of cumulativeness of technological knowledge, and characteristics of the relevant knowledge base (Carlsson *et al.*, 2002). The specific constitution of the industry – actors, institutions – creates the conditions for innovation processes to take place (Malerba, 2002, 2005). So a deep understanding of how innovation works and can be improved, can be achieved by studying innovation on an industry level, and better still on the level of specific projects, as we do in our study by focusing on product innovation the F&B industry.

Management of the innovation process

The management of the product innovation process covers a number of stages. A simple product development model distinguishes three stages: from the idea stage through development to final product launch; others are more refined. According to Wheelwright and Clark (1992) management of the product innovation process is the management of the 'development funnel': a gradual process of reducing uncertainty through a series of problem-solving stages, moving through the phases of scanning and selecting and into implementation, thereby linking market and technology-related resources in the process. Taking decisions in all relevant stages of the product development process on closing down or continuing the development project can best be done using more or less formalised and structured development systems, with structured decisions points. Cooper (1993, 1994) developed the stage-gate model (with six stages) that focuses on systematic screening, monitoring and progressing frameworks (the 'gates') before a next stage in the innovation process is entered.

Routines and resources

Managerial and organisational questions deal with how the company's resources can best be managed: control over scarce resources is the source of economic profits (Teece and Pisano, 1994). The resource concept has been defined as part of the resource-based theory of the firm (Teece, 1982; Wernerfelt, 1984; Peteraf, 1993) which has its roots in the work of Penrose (1959) and Rubin (1973). The theory argues that companies can best be viewed as a collection of resources. The resource-based view of the firm starts from the company's specific strategies for exploiting these specific assets. It deals with how a company's competitive advantage is achieved and sustained. Competitive advantage in this view rests with the company's idiosyncratic and difficult-to-imitate internal resources. Resources can be defined as those (tangible and intangible) assets that are tied semi-permanently to the company (Wernerfelt, 1984). They include product design and production techniques, knowledge of specific markets or user needs, brand names, decision-making techniques or management systems and complex networks for handling markets and distribution sources (Mowery *et al.*, 1998). The more tacit the company's resources, the more difficult for its competitors to copy and use them. In its initial stage the resource-based view focused only on internal resources of innovation; in later stages it also considered inter-organisational co-operations to get access to external resources

(Miotti and Sachweiler, 2003; Mowery *et al.*, 1998; Eisenhardt and Schoonhoven, 1996): these are addressed in the Section 1.1.3.

In the resource-based view, resources are defined not only on the strategic but also on the operational level. The company's resources include all those assets that enable the company to formulate and implement its strategies. On the operational level – that of the management of the production process – resources are the skills, competences and routines that contribute to the implementation of the innovation strategy. Companies use specific managerial and organisational processes that lead to the development and marketing of new products. These managerial and organisational processes are about the way things are done. This can be referred to as the company's 'routines', or patterns of current practices and are related to management style and company culture (Teece and Pisano, 1994).

Technology-related and market-related resources

An important focus of this book is on the role of technology versus market-related resources as they are the main enablers of the innovation process. Technology-related resources can be actors (such as research organisations, suppliers of ingredients or technical equipment) or knowledge-carriers (such as scientific publications, patents) from inside and outside the company that provide input with a technological character; this can be knowledge *pur sang* or embedded in products, processes or services. Similarly, market-related resources, such as market knowledge, brand names, customer knowledge or downstream actors in the distribution channel provide market-related input into the innovation process. We investigate the impact of these resources on the product's short- and long-term market performance: in Chapter 2 we focus on the in-house resources and in Chapter 5 on the external resources in the product-related innovation network (open innovation).

1.1.2 Innovation strategy

The company's strategy was first introduced by Chandler (1962) in his book on 'Strategy and structure'. He was the first to recognise the importance of coordinating various aspects of management under one encompassing strategy. Building on Selznick (1975) who introduced the idea of matching an organisation's internal factors to its external environment, Chandler showed that the formulation of a long-term coordinated strategy was necessary in order to give a company structure, direction and focus of its activities. Mintzberg and Quinn (1991) concluded that the strategy process was rather unpredictable. They positioned themselves opposite the prescriptive school of strategic management, with Michael Porter (1980) as one of its well-known representatives. Their approach deals with how firms should formulate their strategy and not how they actually formulate their strategy. Mintzberg and Quinn (1991) argue that as firms have only limited and imperfect knowledge of their environment and its own strengths and weaknesses, it must take deliberate steps towards its stated objective, measure and evaluate these steps, adjust if necessary and decide on the next steps. So strategy formation is an incremental process in which step-by-step – in accordance with how the organisation learns – changes in strategy formation are made. The concept of eminent strategies acknowledges the ability of the company to experiment and learn and goes back to Weick's (1979) ecological model of variation, selection and confirmation. Companies first act, than evaluate what works well and why and finally select what is preferred. Teece and Pisano (1994) have integrated the three dimensions of strategy formation – competitive markets, firm specific technology and organisation – into the dynamic capabilities approach which includes both the dynamic chance in company strategies and corporate learning (Tidd *et al.*, 2005). This 'learning' approach to strategy development relates to evolutionary theories (as developed by Nelson and Winter, 1982) and empirically-based theories (pioneered by Rosenberg, 1976) of the innovation process.

A company's product innovation strategy is part of the corporate strategy and deals specifically with the company's choice to develop a specific product for a specific market. Following Tidd *et al.* (2005) important inputs for developing a company's product innovation strategy are technological and market opportunities, the company's competences (Hamel and Prahalad, 1994) and the fit with the overall strategy of the company (Hill, 1993). Miles and Snow (1978) have developed a theoretical framework that deals with the alternative ways companies adapt to their external environments by defining their product-market domains and constructing mechanisms to pursue these strategies. Miles and Snow introduced four different types of generic strategies that differ in their adaptive patterns; each type is consistent in its behaviour across the cycle. Three of them - the Prospector, the Analyser and the Defender – represent successful strategic archetypes; the fourth – the Reactor – is unsuccessful (i.e. low performer). The Miles and Snow typology will be used to analyse the relationship between product innovation strategy and product market performance, in Chapter 4.

1.1.3 Innovation network

Innovation networks can be defined as sets of alliances between two or more organisations that are in an interactive way involved in an innovation process. Empirically they are loose or contractual links between two or more companies and other organisations and have a core with both weak and strong ties among constituent members that remain independent agents. Following Imai and Baba (1989) interaction in innovation networks includes three dimensions: between users and suppliers, between R&D, manufacturing and marketing and between physical products, software and services.

Alliances and other collaborative mechanisms such as venture capital investments, outsourcing of specific activities, or licensing-in are considered as devices to gain access to these resources (Kogut, 1988; Hamel, 1991; Hagedoorn, 2002). Following the resource-based view (see Section 1.1.1) alliance formation is driven by a logic of strategic source needs. It is relevant in situations where companies are taking a new and vulnerable strategic position (such new markets, many competitors and pioneering technologies which is an expensive and risky strategy) for which alliances can provide the additional resources to compete effectively (Eisenhardt and Schoonhoven, 1996). Alliance formation can provide critical resources, both concrete ones such as specific skills and financial resources (e.g. Pisano and Teece, 1989) as well as more abstract ones such as legitimacy and market power (e.g. Hagendoorn, 1993; Combs and Link, 2003). The strategic rationale becomes relevant when companies decide to enter into R&D relationships that are not related to their core activities (Teece, 1986) or when new high-risk areas of R&D are entered (Hagedoorn, 2002). Alliances can also improve the market power of a company (together enter new markets, prevent others from entering a market) by providing manufacturing or customer information (e.g. Hamel *et al.*, 1989).

Most studies on innovation networks focus on the company level: the group of private and public actors that are involved in the company's innovation processes. As the focus of our study is on product innovation, we study the product-related innovation network, i.e. the group of external actors that are involved in the innovation process of a specific product. The study in Chapter 5 investigates how the openness of the product-related innovation network relates to the product's short- and long-term market performance.

1.2 Innovation in the Dutch food and beverage industry

The food and beverage (F&B) industry comprises of companies primarily engaged in the preparing, processing, preserving and marketing of F&B products for human consumption. In this book we focus on product innovation in the Dutch F&B industry.

In 2005 the Dutch F&B industry comprised about 4,600 firms that employed 152,800 persons and generated a total turnover of €57.7 billion (LEI, 2008). Most companies in the F&B industry are very small: in 2005 about 40% of these companies employed one single person. About 5% (250) of the total number of Dutch F&B firms have 100 or more employees, including five very large firms with over 2,000 employees. These firms contribute to 77% of the total turnover and to 56% of the total employment of the Dutch F&B industry's (*ibid.*).

Table 1.1 presents data on the economic characteristics of the Dutch F&B industry and its contribution to the Dutch industry as a whole. The table shows that in the period 2002-2005 the contribution of the turnover of the Dutch F&B companies with 100 or more employees to the total Dutch industry's turnover decreased after 2003. In 2005, the highest contribution was generated by the meat (12%) and the fat and oil sector (11%). Other important sectors include beverages (8%), cocoa (7%) and vegetables and fruits (6%). Small contributors are bakery (4%) and grain mill (3%) (LEI, 2008). The decrease in the period 2003 to 2005 was mainly due to the fat and oil sector which fell in this period from a contribution of 14% to 11%. Data on the dairy sector are not available.

The absolute employment figures also decreased during the period 2003 to 2005, but not the F&B industry's relative contribution to the Dutch industry's employment, which stayed

Table 1.1. Economic characteristics of the F&B industry (firms with 100 or more employees).

	2002	2003	2004	2005
Net turnover ($\times 10^9$ Euros)	42.4	42.3	41.7	41.7
Contribution to Dutch industry (% net turnover)	26.6	27.2	24.9	22.4
Employment ($\times$ 1000 persons)	94.5	94.9	89.0	86.2
Contribution to Dutch industry (% employment)	19.2	20.1	19.7	20.0
Turnover made abroad (% of net turnover)	41	46	45	45

(Source: LEI, 2005, 2006, 2007, 2008).

more or less on the same level. This industry is the biggest in the Netherlands: around 1 in 5 employees in the Dutch industry work in the F&B industry. It is also the largest industry when considering its size in terms of value added (*ibid.*).

The Dutch F&B industry has a strong international orientation. First of all, the figures in Table 1.2 on turnover show that about 45% of the turnover made by the larger F&B companies is generated abroad. The table shows for 2007, that the nine largest firms in the Dutch F&B only realised a small part of their turnover in the Netherlands (LEI, 2008). The export figures also illustrate the strong international orientation of the Dutch F&B industry. In 2006 the Netherlands was even the European country with the largest intra-EU export figures, with about € 27.5 billion (CIAA, 2007), with dairy products as the most important export product group.

Third, also the huge investments of Dutch F&B companies in foreign F&B companies illustrate the international focus of the Dutch industry: in 2006 this was more than € 35 billion. Most was invested in US companies: €6.7 billion (LEI, 2008).

Innovation in the Dutch F&B industry

The innovation process in the Dutch F&B industry, as in many other countries, has gone through some major changes during the last decades. Until the 1980's, innovation processes in the F&B industry were mainly directed at improving the efficiency of transforming agricultural products (such as sugar beet or milk) into consumer products (such as sugar, cheese, butter and yoghurt). Since the last decennia a shift has taken place from a production-driven to a (more) market-driven industry which has drastically changed the character of innovation in the F&B industry (and the related agricultural sector). The supply side of the F&B production chain (i.e. agricultural products) was no longer pushing innovation processes (more efficient processing), but the demand side became more important with retailers and through them the consumers as key actors (Van Otterloo, 2000). Important key drivers of this reversal of

Table 1.2. Turnover figures of the largest Dutch F&B companies (x 10^9 Euros, 2007).

Company name	Total turnover worldwide	Turnover of the Netherlands	Category of products
Unilever	40,187	1,135	Food products (and others)
Heineken	12,564	N.A.	Beverages
FrieslandCampina[1]	9,107	2,717	Dairy
Vion	7,140	918	Meat
Nutreco	4,021	565	Animal and fish feed, meat
CSM	2,486	134	Sugar, bakery ingredients, lactic acid
Cosun	1,713	N.A.	Food products, ingredients
Royal Wessanen	1,578	146	Organic food products, snacks

N.A. = Not Available.
[1] Friesland Foods and Campina merged in December 2008 into FrieslandCampina: figures presented for each company separately in LEI (2008) – have been combined in this table.
(Source: LEI, 2008).

the food production chain were the increase in disposable income, the globalisation of food production and distribution and new scientific and technological developments (Costa, 2003). Market orientation and customer consultation as sources of information became crucial for the innovation process; incorporating their interest is a factor for successful innovation for the F&B industry (Kristensen et al., 1998; Omta et al., 2003; Batterink et al., 2006). Product innovations became relatively more important for the F&B industry than process innovations (Arundel et al., 1995). This was recently confirmed by Robertson and Patel (2007) who showed on the basis of patent statistics that from the period 1981-1990 to the period 1991-2000 large F&B companies had moved their technological focus from more process-related fields to more product-related fields.

Notwithstanding the growing importance of market factors and the relatively low R&D intensity of the F&B industry, technology plays an important role. For instance, biotechnology and ICT have become important technological fields in the food production processes of today (Lagnevik et al., 2004). Industries have been classified according to their R&D intensity in classes; the low tech class has a R&D intensity which is below 1%, that of medium-low tech ranges between 1 and 2.5% (Hatzichronoglou, 1997).[1] Firms in low and medium-tech industries, such as the F&B industry, can buy technological expertise from other companies

---

[1] This value is for the indicator 'R&D, related to production' and includes both direct (i.e. in-house) R&D and indirect R&D (R&D expenditure embodied in intermediates and capital goods purchased on the domestic market or imported). For the indicator 'R&D related to the value added' the F&B industry belongs to the group of medium-low tech industries (Hatzichronoglou, 1997).

or public research organisations. They are often important customers of high-tech innovators and can function as lead users who generate ideas and solutions that are tightly targeted at the specific applications in their industries (Von Hippel, 1988). Adaptations of the equipment to the specific needs of the company are often necessary; even when a machine tool is replaced by a newer model, a variety of changes in other related components is frequently necessary (Rosenberg, 1963). Last but not least, low-, medium- and high-technology industries consist of a considerable mix of low-, medium- and high-tech companies (Kirner *et al.*, 2009).

In case of the Dutch F&B industry figures show that this industry has a relatively high level of R&D investment. The R&D intensity in terms of the R&D expenditure as a percentage of the industry's output was relatively high compared to the F&B industry of other European countries over the period 1995-2002. In 2002 it was even the highest: 0.6% (ranging between 0% and 2.4%) against a European average of 0.24% (CIAA, 2006). Also, worldwide the Dutch F&B industry belongs to the Top 3; R&D statistics showed that only the Japanese and Danish F&B companies spend more on R&D (as a percentage of the total value of F&B products); the level of innovativeness is also way above the average (OECD, 2007).

In addition, the innovation performance of the Dutch F&B industry is relatively strong: compared to other industries in the Netherlands, in 2009 the F&B industry belongs to the group of industries that was the most willing to invest (NRC Focus, 2009). SMEs in this industry are also very active in innovation compared to the average of all Dutch industrial SMEs. An EIM study (2008) showed that they have more product innovations (54% versus 34%) and more process innovations (29% versus 16%). And within Europe, the Dutch F&B industry is an outperformer, when compared to F&B industries in other European countries. A recent study (INNOVA, 2008) measured the innovation performance of a number of industries through a combination of three indicators: an index of patenting activity (total number of patent applications per employees at the European Patent office), total factor productivity (combining data for value-added at constant prices, number of hours worked and value of capital stock at constant prices) and an index of market advantage (total export per employee). The results showed that for the F&B industry the Netherlands takes the lead: it has the highest level of innovative activity of all EU25 countries. The relative high performance in innovation of the Dutch F&B industry also could explain the high productivity of the Dutch F&B industry. While the Dutch F&B companies represent only 1.8% of the total number of F&B companies in EU25, their contribution to the European (EU25) total turnover in this industry was 6.3% (INNOVA, 2008).

Within the F&B there are considerable differences in innovation between product groups. The most innovative product group are the dairy products; the share of product innovations in this group as percentage of the total number of product innovations is abour 10% and showed a small decrease from 10.0% in 2004 to 9.2% in 2007. On the second position are the soft drinks (8.2% in 2004 and 7.2% in 2007) and third the frozen foods (showing a steep increase from 5.6% in 2004 to 7.2% in 2007). The least innovative product group were meat products, with a rather constant innovation rate during the period 2004-2007 of around 3% (CIAA, 2005, 2008).

1.3 Research methods

The present study employs a product-oriented empirical approach. Basically there are two approaches in quantitative innovation studies: macro- or meso-economic studies that focus on the relation between national or sectoral economic growth and innovation determinants (such as R&D spending, number of PhDs) and business studies that focus on the determinants of successful innovation at the company level. Within the latter category two different approaches exist: the subject and the object approach. The subject approach focuses on the innovator, Schumpeter's subject of industrial renewal. Studies that follow this approach deal with characteristics of the company and its innovation projects and procedures. The object approach in contrast, collects data on the level of individual innovations and produces information that is directly related to the innovation itself: Schumpeter's object of industrial renewal.

Within the object approach two different methods can be used to identify new products that are introduced to the market. The first uses experts' opinion to identify the innovations, the second identifies innovations through a systematic review of professional technical and trade journals. These journals can be considered as relevant sources for the identification of innovations, as it is the decision of the journal editors – an independent, qualified panel – to include the new product introductions in the 'new product announcements' sections of these journals and not the decision of a stake-holding actor such as the commercialising company (Palmberg et al., 1999). From a statistical point of view the method has advantages as well as the relevant journals constitute a clearly defined population, the journals are edited and the mention of an innovation in the journal implies some kind of a judgement of experts in the field. After identification, additional data on the innovations can be collected through surveys directed at the commercialising firms or from publicly available sources.

In the present study, by screening professional F&B journals published in the second half of 1998, 200 newly announced F&B products could be identified. In order to assure that the products had successfully survived the critical first year on the market in 2000, data were collected on the innovation process, strategy, internal and external resourcing, product innovativeness and market performance through phone interviews based on a structured questionnaire with the managers that had been involved in developing and marketing the products. In the end, complete datasets could be collected for 129 products.

Many studies that investigate the key factors for successful market performance of products only focus on the products' performance shortly after market introduction. There are hardly any data on the key factors for the long-term market performance. This study tries to fill that gap by including an indicator for long-term market performance: at the end of 2005 the long-term market performance of all 129 products was measured.

In our study we draw on an original database of products that have reached the market introduction phase. Measurement of data on products that have been introduced on the market is a direct measure of the commercialisation of the product innovations. This discards innovations that fail prior to market introduction and allows us to study the impact of various factors on the short- and long-term market performance of these products. See Figure 1.1 for an overview of the methods of data collection. We will elaborate on the methods below.

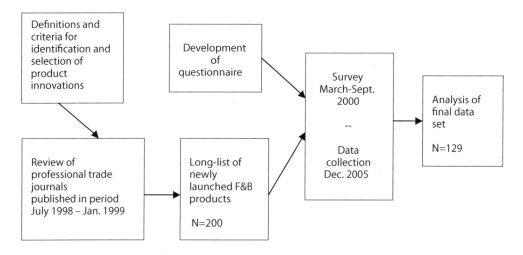

Figure 1.1. Methods used in the present study.

1.3.1 Selection of newly launched F&B products

The set of journals that was systematically reviewed for identification of newly launched F&B products included the Dutch trade journals that cover the whole F&B industry, and journals that cover specific sectors of the F&B industry. These journals have a 'new product' section and include articles that focus on new products. We consulted one of the editors of the main journals (VMT) in order to check the list of journals that we had selected and to provide us with information about missing journals on our list. The final list of journals is presented in Table 1.3. We reviewed the issues published in the period July 1998 to January 1999.

Table 1.3. Professional and trade journals for the Dutch F&B industry used for product identification.

- Bakkerswereld	- Food Management	- VMT - Voedingsmiddelentechnologie
- Catering Magazine	- Food Press	- Voeding en Keuken
- Eyes on Food	- Grootkeuken	- Zuivelzicht
- Food Magazine	- Groot Consument	

We included almost all product categories belonging to the groups 'Manufacture of food products' and 'Manufacture of beverages' covered by Sections 10 and 11 of the NACE Rev.2 classification (Nomenclature statistique des activités économiques dans la Communauté Européenne). However, we have not included meat and fish products (10.1, 10.2) because of the very low innovativeness of this sector (Batterink *et al.*, 2006; INNOVA, 2008) except for snacks and ready-to-eat meals that contain processed meat or fish. The snacks and prepared meals belong to the sector 'Manufacture of other food' (NACE code 10.8). We also excluded prepared animal feed (10.9) as the study is focused on food and beverages for human consumption.

We included product announcements for the consumer market, the food service market and the industrial market (i.e. ingredients). As we were only interested in product innovations of the Dutch F&B industry, we only included products that had been developed and commercialised in the Netherlands. We did not add criteria for the level of innovativeness of the product. Journals were also cross-checked against one another in order to avoid double entries of new products. In total we identified 200 newly launched products. For each product, information was gathered on the product's name and the name of the company that had developed the product.

1.3.2 Questionnaire, survey and interviews

Data were collected by means of a survey using a structured questionnaire. After having tested the draft questionnaire through ten pilot interviews in January 2000 with managers that had been directly involved in the product development process and the redrafting of the questionnaire, data collection took place in the period March to September 2000. The data were gathered from phone interviews with managers that had been directly involved in the F&B product development process. Data on long-term market performance were collected in December 2005, seven years after product announcement, through telephone interviews with the companies asking if the product was still on the market. See the Appendix for the questions and related variables.

The group of 129 products included 76 products for the consumer market, 37 for the industrial market and 16 for the food services market. The products had been produced by 66 companies. The 71 other products could not be included for several reasons: they were either not developed in the Netherlands or not brought to the market, withdrawn from the market, not all data could be collected, the responsible managers could not be identified or contacted or the company refused to cooperate.

1.4 Research questions and hypotheses

The overall objective of this book is *to analyse the key factors for successful product innovation in the F&B industry*. In order to meet this objective we have formulated four research questions, which we will try to answer in Chapters 2 to 5. The study in Chapter 2 focuses on the product innovation process and aims to understand how differences in the product development processes, especially with regard to technology- and market-related resources relate to the short- and long-term market performance of new versus improved products. The study in Chapter 3 explores the differences between the innovation processes of products for the consumer and for the industrial market. The study in Chapter 4 investigates the relationship between product innovation strategy and product market performance and the study in Chapter 5 focuses on the openness and composition of innovation networks of new and improved products and how the use of specific technology- and market-related external resources relate to the short- and long-term market performance of these products.

This section introduces the four different parts of the study and presents the specific research questions and the hypotheses that have been tested in the different chapters.

1.4.1 Innovation process I: new versus improved products

The first study presented in Chapter 2 focuses on the management of the product innovation process. This is about the company's routines, or patterns of current practices (Teece and Pisano, 1994). There is general agreement that the development of new products can best be organised as a structured process, with clear decision points and agreed rules on the basis of which go/no-go decisions can be made. The phased development model includes several stages; from the first idea stage to the last stage which is the product's market introduction (Cooper, 1980, 1983, 1985; Cooper and Kleinschmidt, 1987), thereby linking market-related and technology-related streams (Tidd *et al.*, 2005). The early stage (that of idea generation) is particularly important as it provides the objectives for the next stage: the development process. Investments in high-quality execution of predevelopment activities are repaid in reduced development times and improved success rates, higher profitability and larger market share (Cooper, 1993; Cooper and Kleinschmidt, 1987, 1990)

Only a few studies on product innovation address the differences in the management of new versus improved product development processes. Earle *et al.* (2001) found that in the case of new products, product development involves relatively more market-related activities (i.e. activities such as market research, consumer tests, etc.) because the market is still unknown. In the case of improved products, the market is known and only adaptations are needed. On the other hand, Balanchandra and Friar (1997) argue that for new products the market still has to be conquered; therefore technology-related activities (i.e. dealing with the technical aspects of the product and its production process) are more important. For the improved product, once it has reached a certain market share, market factors are very important in order to keep and increase that market share (*ibid.*).

There is no consensus in literature on the importance of market- versus technology-related resources in the innovation process of new versus improved products. Therefore the study presented in Chapter 2 has an explorative character; its main aim is to investigate for the group of new products and improved products separately the differences in their product innovation processes and how the technology-related and market-related resources used in these processes relate to the performance of these products on the market, in the short and long term.

The study in Chapter 2 focuses on the following research question:

Research question 1 (RQ1): *Which technology- and market-related resources play a key role in the short- and long-term market performance of new versus improved products?*

In tackling this question, we have selected a number of innovation process determinants. First are the so-called upfront activities, predevelopment activities leading to a clear set of objectives for the product development process (Cooper, 1993). Second are a number of routines, company practices and processes that have shown to be successful for optimising the innovation process. One is the involvement of technology- and market-related company functions in the product innovation process and more specific in cross-functional teams that have autonomy and are responsible for the whole trajectory from idea generation until final commercialisation of the new product. The use of cross-functional teams is an important factor for success (e.g. Cooper, 1983; Wheelwright and Clark, 1992; Calantone *et al.*, 2002). Another is the evaluation of the innovation process that enables the company to distinguish between successful and less successful projects and to draw lessons for new projects (Mansfield and Wagner, 1975). These organisational routines are the visible artefacts of the company's innovation culture; invisible are the patterns of shared values, beliefs and agreed norms which shape the behaviour and give meaning to the activities of the company; they keep the company together (Peters and Waterman, 1982). For that reason we included both organisational and company culture aspects in the study.

1.4.2 Innovation process II: consumer versus industrial products

Having explored the specificities of the product innovation process of new versus improved products, Chapter 3 studies the characteristics of different types of product innovations in the F&B industry: consumer versus industrial products. Earlier studies on product innovation in this industry focused only on products for the consumer market or covered all products but did not distinguish between consumer and industrial products. Both consumer and industrial products are products of the F&B industry, but may be different in their innovation processes. Consequently innovation management may have to focus on different aspects. The descriptive study presented in this chapter investigates the role of technology-related and market-related resources in the innovation process of the two product groups, in order to answer the following research question.

Research question 2 (RQ2): *What are the differences in the involvement of technology-related and market-related resources in the innovation process of consumer versus industrial products and in their short- and long-term market performance?*

The development of ingredients for functional foods, but also other ingredients, such as those used in bakery mixes (enzymes, yeasts) is relatively knowledge intensive. Research is needed in order to investigate the functionality of the specific ingredient and to develop formulations for optimal activity (Enzing and Van der Giessen, 2005). Considering this, it can be expected that in the development of F&B ingredients technology-related resources are used relatively more often as compared to consumer products. The higher knowledge intensity of the development of ingredients as compared to consumer products implies that more investments have to be made. As a consequence a longer time period is needed to earn back the investments. So it can be expected that industrial products will be developed to be on the market for a longer period of time. For consumer products we expect that they will perform better in the short term and will be on the market for less time, so will perform less well than industrial products in the long term. We expect this because of the need for regular new product revisions in order to catch the consumer's eye. On the basis of the above argumentation we have formulated the following two hypotheses:

Hypothesis 2a (H2a): *The innovation process of consumer products will include more market-related resources, whereas that of industrial products will include more technology-related resources.*

Hypothesis 2b (H2b): *Consumer products will show better short-term market performance, whereas industrial products will show better long-term market performance.*

1.4.3 Innovation strategy: prospector versus analyser or defender strategy

Chapter 4 focuses on the third research question addressed in this book.

Research question 3 (RQ3): *What is the impact of the product innovation strategy on the product's short- and long-term market performance?*

Empirical investigations have shown that an articulated innovation strategy is an important factor for the innovative product's success on the market. A strategy provides guidelines for questions such as which market to enter, with which products and which skills to develop (Lester, 1998). As companies in the F&B industry increasingly have to compete on the basis of new and more advanced products, developing a product innovation strategy is becoming increasingly important. Earlier studies on strategies of F&B companies mainly deal with the extent of geographical market coverage and differentiation, entry barriers and bargaining power (McGee and Segal-Horn, 1992; Hyvönen, 1993). More recent studies also address innovation as a central aspect of F&B companies' strategies (Gilpin and Traill, 1999; Martinez and Briz, 2000; Traill and Meulenberg, 2002, Avermaete *et al.*, 2004; Batterink *et al.*, 2006). The study presented in Chapter 4 goes a step further by investigating how the innovation strategies of F&B companies are related to their outputs: the product's short- and long-term market performance.

Miles and Snow (1978) have developed a theoretical framework of strategic archetypes which distinguishes between three types: prospector, analyser and defender strategies. They represent alternative ways in which companies can define their strategies and construct mechanisms to pursue these strategies. Each type shows a consistent pattern of response to the changing environments in which they operate. They have their own unique market strategy and a particular configuration concerning technology, structure and organisation that is consistent with this market strategy. The importance of product innovation differs per strategy type and consequently so does the need for (new) resources. Companies following a prospector strategy foster growth by developing new products and exploiting new market opportunities. Those following a defender strategy grow mainly through market penetration and by producing and distributing their goods as efficiently as possible. Analyser companies combine the strengths of both the prospector and the defender strategies; they attempt to minimise risks by moving towards new product and markets only after their viability has been demonstrated, while maximising the opportunity for profit by its stable product and market areas (Miles *et al.*, 1978).

The F&B industry can be characterised as a scale-intensive industry, where process innovation is the main mechanism for more efficient and effective production (Pavitt, 1984)[2]. Companies in this industry generally turn to suppliers to purchase specific instruments, specialised machinery, materials and related services. Indeed, through this mechanism F&B companies profit from new technological developments in upstream sectors (Klevorick *et al.*, 1995; Rama, 1996; Christensen *et al.*, 1996; Martinez and Briz, 2000; Traill and Meulenberg, 2002). For that reason one would expect that the defender strategy is the most common product strategy type in the F&B industry. However, it can be expected that the recent changes in this industry will have had its impact on the companies' innovation strategies. We refer in particular to the reversal of the dynamics in the agrifood chain towards more demand-driven (i.e. consumer preferences and food choice in combination with the growing role of retailers). Therefore it can be expected that F&B companies will increasingly apply prospector-type strategies by developing more new products. When these new products will catch on, they will give long-term commercial benefits in terms of sales and overall profitability (Joppen, 2004); many studies found that innovative products are more successful than 'me-too' products (Van Trijp and Meulenberg, 1996; Hoban, 1998; ECR Europe, 1999; Knox *et al.*, 2001). These new products will provide prospector companies with a competitive advantage. On the basis of this argumentation we have formulated Hypothesis 3 which will be tested in Chapter 4:

---

[2] Pavitt (1984) found that sectors differ in the patterns of technological change and grouped them into four categories; supplier-dominated, scale-intensive, specialised-supplier and science-based industries.

Hypothesis 3 (H3): *The products of F&B companies that intend to follow a prospector innovation strategy will show better long-term market performance than the products of companies that intend to follow another (analyser, defender) innovation strategy.*

1.4.4. Innovation network: open innovation

Innovative companies increasingly use resources from outside the company to speed up the innovation process. Chesbrough (2003) introduced the concept of 'open innovation' to describe this phenomenon. He argues that the model of 'closed innovation' which was very dominant in 1980's has been opened and companies increasingly consider external sources of knowledge, competencies and creativity as important inputs for their innovation process. Open innovation takes place in networks and is characterised by specialisation, cooperation and knowledge sharing (Chesbrough, 2003). External innovative ideas and the use of external resources to develop these ideas into successful products are considered as important as internal ones. The building and maintaining of external networks has become vital to an innovative company's strategy for survival and growth. This not only applies to R&D-intensive firms; less R&D-intensive firms also rely on external resources as their infrastructure is not sufficient to innovate on their own. However, most empirical studies on open innovation deal with high-tech industries such as biopharmaceuticals, ICT and computers (e.g. Christensen *et al.*, 2005; Fetterhoff and Voelkel, 2006; Dittrich and Duysters, 2007) and with a strong focus on large and predominantly US-based firms (Chesbrough, 2003, 2006). Empirical investigations into open innovation in low-tech industries, such as the F&B industry are scarce (Sarkar and Costa, 2008). The fact that open innovation might be very interesting to companies in the F&B industry is indicated by the finding of Archibugi *et al.* (1991) that F&B firms rely even more on external resources than the average for all industries. The study in Chapter 5 tries to fill this gap. It focuses on how the openness and composition of the product-related innovation network relate to the short- and long-term market performance of the newly introduced product. The research question that is addressed in this study is:

Research question 4 (RQ4): *What is the impact of the level of openness and the composition of the product-related innovation network on the product's short- and long-term market performance?*

In the F&B industry external resourcing takes place through the interface in the food value chain; the typical interdependencies in this chain are of significant importance for the innovation process (Boon, 2001). Additionally, resources provided through horizontal relations, such as companies supplying machinery and equipment, consultants proving marketing or recipe advice and research organisations can be used. We expect that F&B companies that innovate in networks are more successful than companies that don't and have hypothesised this as follows:

Hypothesis 4 (H4): *The more open the product-related innovation network, the better the product's short- and long-term market performance.*

The innovation networks can involve both technology-related and market-related actors. As F&B companies have only limited resources for scientific and technological activities, they are dependent on external knowledge inputs when developing more innovative products. This input from technology-related actors can come directly from research organisations (universities, research institutes, polytechnics), and other companies and also indirectly by purchasing goods (such equipment, machinery, ingredients, raw materials) in which new scientific and technological developments are embodied. We investigate the composition of the innovation networks of new versus improved products and expect that the role of external technology-related actors will be more crucial for the successful market performance of new products in the short and long term. For improved products we expect the involvement of technology-related actors to be crucial only for the short term. Market-related actors are expected to be specifically crucial for successful market performance of new products. For example, they provide market knowledge and advice based on extensive market research and advanced and prepared marketing and brand positioning efforts. Improved products already have a market, and external input from market-related actors is less crucial for the product's market performance.

Finally, in Chapter 6, a synopsis of the key findings and their theoretical implications is given. The chapter then draws conclusions about the contributions of the study to literature and provides suggestions for further research.

2. Innovation process I: new versus improved products

2.1 Introduction

Chapter 2 presents an explorative study on the differences in the innovation process of new versus improved F&B products. The study was set up to analyse how the involvement of technology- and market-related resources relate to the short- and long-term market performance of these products.

There is no consensus in literature about what is most important in the innovation process of new products versus that of improved products: technology or market. On the basis of Earle *et al.* (2001) it could be argued that in the case of new products, market-related activities (i.e. market research, consumer testing) are more important because the potential market (segments, share) is still unknown and has to be explored. In the case of improved products, the market is known and only adaptations are needed; so market-related activities are less crucial for the improved product's market performance (*ibid.*). However, on the basis of Balanchandra and Friar (1997) one could also propose the opposite. Since the market still has to be conquered for new products, technology-related activities (i.e. dealing with the technical aspects of the product and its production process) are more important, because a good quality product will always find a market (see also Link, 1987; Rource and Keeley, 1990; Calantone *et al.*, 1993). For the improved version of a new product, market factors are more important for keeping in touch with changing consumer wishes in that market (Balanchandra and Friar, 1997). On organisational aspects there seems to be some agreement. Earle *et al.* (2001) argue that improved products can be developed using standard procedures while new products require less standardised activities, because flexibility and creativity are prerequisites and problems cannot be predicted. Moreover, Balanchandra and Friar (1997) claim that organisational factors are most important for incremental product innovations.

The discussion on the importance of market- and technology-related resources is particularly interesting for the F&B industry; the current emphasis on market orientation (Grunert *et al.*, 2008) especially in marketing literature, against the current growth of knowledge-intensive products such as functional foods (see for instance Stein and Rodríguez-Cerezo, 2008). Companies in this industry have to compete on the basis of regular launches of new and improved products on the market. However, the competitive position of companies in the F&B industry is under pressure (Wijnands *et al.*, 2006). A major problem for this industry is the high failure rates. A large European study concluded that two in every three newly introduced consumer products fails within the first year after product launch. One year after product launch between 20% (for innovative products) and 10% (for less innovative products) of the products were still available in 50% of the supermarkets (Van Poppel, 1999). Therefore one of the most important challenges of the management in this industry is to organise the product innovation process in such a way that the product's success rates on the market will increase. Although there is growing recognition for the role of product newness in the product's market performance (Hoban, 1998; ECR Europe, 1999; Lord, 2000), there is still a lack of studies focussing on the specific aspects of the product development process of new versus that of improved products. Both technology- and market-related resources play a key role in food innovation processes (Traill and Grunert, 1997), but for new food products these could be different than for improved products. The main aim of the present study is to fill this gap. We do this by investigating for the groups of new and improved products separately, the differences in their respective innovation processes and how the technology-related and market-related resources involved in these processes relate to the performance of the product groups on the market.

The research question of the study is:

RQ1: *Which technology- and market-related resources play a key role in the short- and long-term market performance of new versus improved products?*

Our study takes a new approach as it does not focus on the performance of a company and thus of all the company's products, but on the performance of a specific product that was in development. In addition, we not only investigate the impact of innovation process-related determinants on the product's performance soon after market launch, but also after several years, because we argue that a sustainable market position is the best condition for a good competitive position.

The study focuses on the Dutch F&B industry. Within the Netherlands, it is the largest manufacturing sector, representing 22.4% of the total turnover of all manufacturing industries (LEI, 2008; data for 2005). It is also the leading employer representing 20% of the total Dutch manufacturing industry (*ibid.*). The Dutch F&B industry is one of the biggest players in Europe: six Dutch companies belong to the Top-15 European F&B companies by sales (CIAA, 2009; 2008 data). The Dutch contribution to the European (EU25) total turnover in this industry was 6.3%, while the Dutch F&B companies only constitute 1.8% of the total number of F&B companies in EU25 (INNOVA, 2008).

In the next section we introduce the theoretical framework. In Section 2.3 the data and methods are presented. Based on an empirical study using data on 129 product introductions on the Dutch F&B market, the results are presented in Section 2.4. The last section draws conclusions and discusses the management implications.

2.2 Theoretical framework

We have developed a conceptual model that integrates upfront activities, organisational routines and innovative culture of the company in the innovation process and relates this to the product's short- and long-term market performance. We argue that the relationship between the product's innovation process and the product's market performance is influenced by the innovativeness of the product. The concepts in the model are presented in Figure 2.1 and will be discussed in more detail in the next three sections (2.2.1 to 2.2.3).

2.2.1 Innovation process

Empirical studies on the organisation and management of innovation, such as SAPPHO, the first study that compared successful and unsuccessful innovations in the chemical and instruments industries (Freeman *et al.*, 1972; Rothwell *et al.*, 1974), NewProd (Cooper, 1979; Cooper and Kleinschmidt, 1987; 1990), the Stanford Innovation Project (Maidique and Zirger, 1984) and many others have identified several characteristics of the innovation process that are related to success or failure. The reader is referred to Balbontin *et al.* (1999) for a comparative analysis of the 'historical' studies and to Van der Panne (2004) for a literature review of 31 empirical and 12 theoretical studies.

There is agreement on the importance of a number of specific activities for successful product innovation. They can be clustered in activities that aim at the identification of opportunities and other preparatory activities in the upfront phase of the innovation process and organisational routines that aim at organising and structuring the product development process. These activities and organisational routines are the visible artefacts of what can be termed the company's innovation culture.

Invisible aspects of the company culture are norms and values that shape how company employees relate to each other, to their work and to the outside world (Van Reenen and

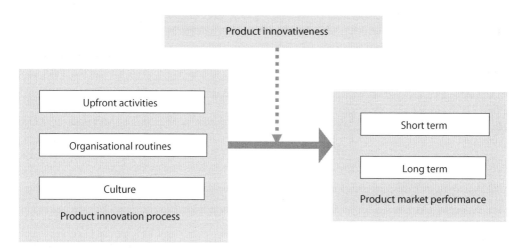

Figure 2.1. Research model.

Waisfisz, 1995) and which reflect 'the way we do things around here' (Tidd *et al.*, 2005). They shape the behaviour and give meaning to the activities and keep the company together (Peters and Waterman, 1982). These three – upfront activities, organisational routines and company culture – are the main determinants of the innovation process we focus on in the study presented in this chapter.

Upfront activities

Companies tend to move directly from a good product idea into the development stage without any in-depth analysis in the predevelopment stage which would help them to better define the product. They ignore these preparatory upfront activities in order to shorten development time to be first on the market. Many F&B products have failed because 'of a rush to the market without shelf-life trails, with disastrous quality results' (Earle *et al.*, 2001). The upfront phase includes the most important activities of the innovation process (Cooper and Kleinschmidt, 1987). In general, investments in high-quality execution of predevelopment activities are repaid in reduced development time and improved success rates, higher profitability and larger market share (Cooper and Kleinschmidt, 1987, 1990). These activities are very helpful for getting a well-defined and sharp product definition. It provides a clear set of objectives for the development process on the basis of which this process can proceed more efficiently (Cooper, 1993).

In our study we focus on market- and technology-related upfront activities. Important upfront market-related activities through which market-related resources get involved in the innovation process are market analysis and market assessment (Freeman *et al.*, 1972; Cooper and Hlafeck, 1975; Cooper, 1980; Hopkins, 1981; Maidique and Zirger, 1984). These can be generic studies on future trends and market surveys that provide information on size, growth and segmentation of the market, specific studies on customers and end-user trends and needs (Von Hippel, 1988; Hoban, 1998; Kristensen *et al.*, 1998; Stewart-Knox and Mitchell, 2003) or testing of the product concept on potential customers (Rudder *et al.*, 2001; Dahan and Hauser, 2001). Important upfront activities through which technology-related resources get involved in the innovation process are, for instance, preliminary technical assessment and checking of patents (Schmidt, 1995; Zirger, 1997). Other upfront activities include the initial screening within the company for commitment of resources (Cooper, 1975; Rudder *et al.*, 2001) and preliminary business and financial assessment (Cooper and Hlafeck, 1975; Mansfield and Wagner, 1975).

Organisational routines

According to Tidd *et al.* (2005: 70) success in innovation depends on two key ingredients: the technical resources and the capabilities in the organisation to manage them. The company's capability to manage the resources relates to its organisational processes, also referred to as 'routines' or patterns of current practice and learning. These processes form the essence of the firm's dynamic capabilities and are the basis for its competitive advantage (Teece *et al.*, 1997: 518). In these 'routines' the particular behaviour of a company has emerged as a result of doing things over a period of time; the current practice is what has appeared to work well. Organisational routines include, for instance, how innovative processes are managed, project teams are selected and their tasks are planned and monitored. They can be described as involving established sequences of actions for undertaking tasks that include a mixture of technologies and methods, formal procedures or strategies, but also informal conventions or habits (Levitt and March, 1988). These routines are constantly evolving when new routines turn out to work better than the old ones. They exist independently of particular personnel; new employees learn them as they enter the company (*ibid.*). They are firm-specific and are what make companies differ from each other (Tidd *et al.*, 2005).

In this study we focus on a number of specific routines: the use of formal product development plans, evaluation procedures, go/no-go decisions and cross-functional teams. Several studies found that the consistent use of a formal and multi-step process is an important key to success (Freeman *et al.*, 1972; Rothwell, 1976; Maidique and Zirger, 1984; Wind and Mahajan, 1988; Lester, 1998; Hoban, 1998). Through a formal process multidisciplinary inputs in the innovation process can be organised, many different activities can be integrated and uncertainties reduced. Specific key activities (such as the upfront activities) can be made mandatory and the quality of its execution can be controlled by introducing check points in the process. An evaluation procedure enables the company to distinguish between successful and unsuccessful projects and to draw lessons and learn from them (Mansfield and Wagner, 1975; Maidique and Zirger, 1984). Companies that use a formal procedure are able to increase the portion of the total R&D budget going to products that are ultimately successful; they show faster new product introductions, less recycling to redo steps and a higher success rate of launched products (Booz *et al.*, 1982; Cooper, 1993). Procedures that institutionalise and thereby facilitate the communication between the different company functions are also important. For instance, input from many different sources is needed, especially from technology-related and market-related company functions (Cooper, 1983; Hoban, 1998; Grunert *et al.*, 2008). Cross-functional teams are a means of integrating the necessary company functions; they are found to be an important success factor (Wheelwright and Clark, 1992; Brown and Eisenhardt, 1995).

Company culture

A company's culture can be defined as the pattern of shared values, beliefs and agreed norms which shape behaviour (Tidd *et al.*, 2005). A company culture that is open to innovation is one of the main preconditions for an innovative company to be successful (Lester, 1998). A company's culture deals with several aspects such as openness to R&D results (Mansfield and Wagner, 1975), flexibility (Liebeskind *et al.*, 1996) and mutual trust. The latter stimulates communication between company functions and thus cooperation (Calantone *et al.*, 1993; Rochford and Rudelius, 1997). An innovative company culture can raise the awareness, creative skills and problem-solving abilities of all employees and lead to higher levels of participation in innovation (Tidd *et al.*, 2005).

2.2.2 Product innovativeness

In literature various classifications of product innovativeness have been proposed. The OECD definition (Oslo Manual) distinguishes between major product innovation (also referred to as radical product innovation) and incremental product innovation (OECD, 2005). Classifications that have been developed for product innovations in the F&B industry use a multi-step approach indicating several stages of innovativeness. Fuller (2005) distinguishes between creative products, innovative products, four types of products that deal with a specific change in the existing products (repackaging, reformulation, repositioning, new form or size) and line extensions.

ECR Europe (1999) and Herrmann (1997) make a distinction between three different types of product innovations: market innovations (real novelties on a market); quasi-new products (innovations that improve the characteristics of existing products) and me-too products (products that are new to the company but not to the market). According to Traill and Grunert (1997) all these classifications boil down to two dimensions: newness to the market and newness from a technological perspective; these correspond to the two most important sources of innovation.

2.2.3 Product performance

Cooper and Kleinschmidt (1987) found three independent dimensions that characterise new product performance: financial performance, opportunity performance and market impact. Others – including Clark (1989) and Brown and Eisenhardt (1995) – argue that measurement of success should include factors such as actual sales and profits, technical aspects, impact on the company's reputation, etc. (see for an overview Murphy *et al.*, 1996). Griffin and Page (1996) investigated what are the best measures for product development success and failure. they found that customer satisfaction and customer acceptance were amongst the most useful customer-based measures for success for several project strategies, but market share was the most useful customer-based measure for projects involving new-to-the-company products or line extensions. On the basis of a survey of 300 food processing companies, Kristensen *et al.* (1998) found that – from a set of eight different success criteria – the impact on the company's market share and on the company's earning capabilities are by far the best indicators of the success of new food products.

Hultink and Robben (1995) have investigated how the company's time perspective influences the importance the company attaches to 16 core measures of a new product's success. For the short term, product-level measures such as speed-to-market and whether the product was launched on time are perceived as important and for the long term, the customer acceptance and financial performance, including attaining goals for profitability, margins, and ROI. Four factors are perceived as being equally important for successful short- and long-term performance: customer satisfaction, customer acceptance, meeting quality guidelines, and product performance level. Tidd *et al.* (2005: 37) also argue that the time perspective needs to be considered: success in the short term 'might be a result of a lucky combination of new ideas and receptive market at the right time' but this is not enough for a company that aims at sustainable growth over a long period of time.

2.3 Data and methods

2.3.1 Data collection

We have collected data on new F&B product introductions by means of a systematic review of the issues of 11 different Dutch food trade and professional journals that were published in the second half of 1998. The innovations were identified by sampling the editorially controlled 'new product announcements' sections of these journals. The advantage of this method is that it has been the decision of the journal editors – an independent, qualified panel – to include them and not the decision of a stake-holding actor such as the commercialising company. We screened the Dutch food trade and professional journals and identified 200 products that had recently been launched.

For each product we gathered information from the journals about the product's name and the name of the company that had developed the product. Additional data on the product's innovativeness, market for the product, on upfront activities (both technology-related and market-related), routines, company culture and the product's market performance were collected by means of a survey, using a structured questionnaire.

For upfront activities we have made a distinction between the very beginning of the product innovation process – the idea stage – and the stage at which detailed preparatory investigations are made in order to define the product and verify its attractiveness prior to heavy spending. The latter stage is what Cooper (1993: 113) refers to as the critical homework stage: 'the one found to be so often weakly handled'. The companies that had launched the products were asked to indicate which sources of ideas they had used for the product. We have distinguished between technology-related and market-related sources. In addition we also included the management of the company as this is an important source of new ideas (Enzing *et al.*, 1996). For the measurement of activities in the 'homework' stage a list of technology- and market-related preparatory and test activities based on Cooper (1993) was used. In addition, we measured the involvement of technology- and market-related company functions in the innovation process. The logistics function was also included as this is very relevant for this industry, especially for those product groups where freshness is an important aspect, such as ready-to-made meals.

We included questions on the use of four different organisational routines: formalised product development plans, evaluation procedures, go/no-go decisions and cross-functional teams. Company culture was measured using seven questions each dealing with one dimension of company culture, based on Sanders and Neuijnen (1987). The respondents were asked to indicate how their company performs on each of the culture dimensions using the scale: low, medium, high.

Product innovativeness was measured using three indicators. First of all the companies were asked: is the product new or is it an improved version of a product that was already on the market? Two other indicators measured the newness of the product attributes and of the knowledge used for developing the product. For newness of product attributes the respondents could indicate for which (combination of) attributes the product was new or improved: raw materials, ingredients, processing, recipe, shelf-life conditions, readiness to use/eat, packaging and/or nutritional value. The value for the ordinal indicator 'Newness of product attributes' is the sum of the scores for each of the attributes: the more new/improved attributes, the higher the product's innovativeness. For the newness of the knowledge the respondents were asked to choose between: has new scientific-technological or other knowledge been used for the development of the product that was developed by the company itself, was it developed by others, or has existing and earlier applied knowledge been used?

The market for the product was measured by asking for which market the product was developed: for the consumer market, the industrial market or the food service market.

The product's market performance was measured twice: one and a half years after product launch announcement and seven years after product launch announcement. Short-term market performance stands for the financial and market impact of the product one and a half years after it was announced in the trade journals. It was measured through a combination of two indicators: impact of the product on the company's market share and on the company's turnover. Both were measured using a 3-point Likert scale (no change, small increase, large increase in market share/company turnover due to product introduction on the market). The value of the indicator 'Short-term market performance' is the average of the scores on the two indicators: low performance stands for no change, medium performance for small increase and high performance for large increase. Long-term market performance stands for the market status of the product seven years after it was announced in the trade journals. It was measured using a dichotomous indicator: the product is still on the market or no longer on the market.

In the period March-September 2000 these data were gathered from phone interviews with the managers of the firms who had been directly involved in the innovation process of the specific products. Data on long-term market performance were collected in December 2005, seven years after product announcement, through telephone interviews with the companies.

Complete data sets have been collected for 129 of the 200 new products. Of the 71 products that are not included 45% were not eligible (not developed in the Netherlands, not brought to the market after all, withdrawn from the market) and for the other 55% data could not be collected for several reasons (responsible person could not be identified or not be contacted, company refused to cooperate). Compared to sending questionnaires by post or email – which have an average response rate of about 30% –, our method has led to a relatively high response rate of 76% (129 from an eligible population of 170). Data collection on long-term market performance could be completed for all 129 cases.

2.3.2 Methods

The relationships in the conceptual model were tested by means of correlation analyses for each of the variables for upfront activities, organisational practices and culture with short- and long-term market performance for the group of new and improved products separately. Regression analyses tested the relative importance of a number of specific technology- and market-related resources used in upfront activities against each other for short- and long-term product performance. Independent variables were variables for these technology- and market-related resources. We used product innovativeness and type of market as control variables. Linear regression analysis was performed for short-term market performance as a dependent variable. The predictive power of the linear regression model was measured by the coefficients of determination (R^2), which is the proportion of the total variance explained by the model. For long-term market performance as a dependent variable, a binary logistic regression analysis was used. The Mann-Whitney test (2-Independent samples test) is suitable for non-parametric data analysis and was used to test between the groups of new and improved products.

2.4 Empirical results

The 129 products represent a large variety of product groups of the F&B industry: ready-to-made or use meals (including breakfast products) and snacks; processed fruit and vegetable products; dairy products; grain mill, starch, bakery and farinaceous products; beverages; chocolate and sugar confectionery and ingredients for food products. Of these 129 products, 61% are new products and 39% are improved products. Most new products are in the ready-to-made product group (80%), the food ingredients (73%) and the grain mill, starch, bakery and farinaceous product group (63%). The smallest number of new products is in the chocolate and sugar confectionery group (40%). In the other three groups (fruit & vegetables, dairy and beverages) the ratio between new and improved products is about fifty-fifty. More than half of the products were for the consumer market (76 products), 37 products were for the industrial market and 16 products for the food services market. The 129 products were produced by 66 different companies. Some companies had several business units or divisions, which are spread across several locations in the Netherlands.

2.4.1 Descriptive results

New products have significantly more new product attributes, compared to improved products (see Table 2.1). The two product groups hardly differ for the 'Newness of knowledge used' variable: new products have been developed with only slightly – but not significantly – newer knowledge.

Upfront activities

In our study, upfront activities include technology-related and market-related (re)sources involved in the idea generation stage and in the preparation and testing stage. In addition we also measured the involvement of technology-related and market-related company functions in the innovation process.

In the idea generation stage, there are only small – and non-significant – differences between new and improved products in the use of the three internal sources of ideas we have included. The company's marketing/sales department were used most frequently (about 90%) as a source of innovation (see Table 2.2). In about two thirds of the cases ideas from the company's R&D department and in about one third of the cases the board of directors of the company was a source of new product ideas. In the preparatory and testing stage three technology-related activities were carried out more often for new products than for improved products: technical feasibility, checking of patents and licences and testing of product qualities. The fourth (test of production) was used in fairly equal measure for both product groups. Two of the four market-related activities – market forecast and consumer testing – were used more often in new product development. In the case of improved products marketing tests were used more often and the product concepts were tested by potential customers. However, the two product groups show only small differences for most of the preparatory and testing activities. They only

Table 2.1. Innovativeness of new and improved products (mean, standard deviation between parentheses).

	New products (N=79)	Improved products (N=50)	DbG[1]
Newness of product attributes [2]	4.06 (0.99)	3.54 (1.09)	***
Newness of knowledge used [3]	1.61 (0.87)	1.58 (0.81)	

[1] DbG - Difference between Groups. Significance of difference between the two product groups (using Mann-Whitney analyses; excluding cases test-by-test; exact sig 1-tailed) is indicated by *** for: P<0.01.
[2] Values range from 0 to 8.
[3] 3-point Likert scale.

Table 2.2. Upfront activities (frequencies).

	New products (N=79)	Improved products (N=50)	DbG[1]
Idea stage			
TR: Sol - company's R&D department	71%	65%	
MR: Sol - company's marketing/sales department	92%	86%	
Other: Sol - management/board of the company	34%	30%	
Product development stage			
TR: technical feasibility	82%	68%	*
TR: check patents and licences	58%	50%	
TR: test product qualities	89%	84%	
TR: test production	91%	92%	
MR: marketing research	89%	84%	
MR: test product concept by potential clients	85%	90%	
MR: test marketing activities	86%	88%	
MR: test product by consumer	71%	54%	*
Company functions involved			
TR: R&D function	100%	95%	*
TR: purchase function	25%	16%	
MR: marketing function	44%	38%	
MR: sales function	35%	30%	
Other: logistics function	14%	4%	*

TR = technology-related; MR = market-related; Sol = Source of ideas.
[1] DbG - Difference between Groups. Significance of difference between the two product groups (using Mann-Whitney analyses; excluding cases test-by-test; exact sig 1-tailed) is indicated by * for $P<0.10$.

differ significantly in the use of technical feasibility and consumer tests; both are carried out more often in the preparatory and testing stage of new products.

All five company functions were involved more often in the case of new product development; two of them – the R&D function (technology-related) and the logistics function – significantly more often (Table 2.2). The latter could be explained by the fact that the group of new products consists of a relatively large number of ready-to-make products.

Organisational routines

Three of the four organisational routines that we included in our study were used more often in the case of new products compared to improved products (Table 2.3). The product groups differ significantly from each other for the use of two of them: the product development plan and the evaluation *ex post*. Cross-functional teams were used more often for improved products.

We created a new variable 'Scale of preparatory and test activities' based on the number of preparatory and test activities being carried out in each case: this included the technology-related and market-related preparatory and test activities mentioned in Table 2.2 plus four other preparatory activities (product definition, financial assessment, legal assessment and product plan) which are not included in Table 2.2. A second new variable 'Number of organisational routines' was based on the number of routines being used. The two product groups differ significantly from each other for both variables: they were used more often in the case of new products (see Table 2.3).

Table 2.3. Use of organisational routines (frequencies; mean with standard deviation between parentheses).

	New products (N=79) Frequency	Improved products (N=50) Frequency	DbG[1]
Product development plan	89%	76%	*
Go/no-go decision	88%	81%	
Evaluation *ex post*	71%	56%	*
Cross-functional team	70%	74%	
	Mean (SD)	Mean (SD)	
Scale of preparatory and test activities [2]	6.49 (2.01)	6.10 (1.33)	***
Number of organisational routines [3]	3.24 (1.00)	2.90 (1.08)	**

SD = Standard Deviation.
[1] DbG - Difference between Groups. Significance of difference between the two product groups (using Mann-Whitney analyses; excluding cases test-by-test; exact sig 1-tailed) is indicated by *** for: $P<0.01$; ** for: $P<0.05$ and * for $P<0.10$.
[2] Scale of this variable is 0 to 8.
[3] Scale of this variable is 0 to 4.

Company culture

The results on company culture show only very small and non-significant differences between the group of new and improved products (see Table 2.4). However, all cultural aspects are evaluated as rather relevant for the companies as they almost all score above 2.5. The 'Human focus' is the weakest and the 'Result focused' the strongest cultural characteristic in the companies that have developed the products.

As a number of companies in our sample have developed and marketed more than two products, these can be both new and improved products and in that case their company culture characteristics are represented in both product groups. This could explain the similarity in company culture aspects of the two groups. We have checked this and found that from the total of 66 companies that produced the 129 products, 32 companies (48%) were represented by only (one or more) new products and 19 companies (29%) by only improved products. Of the other 15 companies with both new and improved products, 13 companies (about 20% of all companies; representing 37% of all products) had a rather even spread of products across both product groups. On the basis of this we can conclude that only part of the cultural similarities can be explained by the representation of companies in both product groups.

Short- and long-term market performance of new and improved products

The results show that nearly two thirds (64%) of the products that were announced in the second half of 1998 were still on the market seven years later (see bottom row in Table 2.5). Products were grouped according to their market performance in the short term (measured one and a half years after product announcement) in three categories: low (13%), medium (49%) and high (38%). One quarter of the products with a low short-term market performance is still

Table 2.4. Company culture (mean with standard deviation between parentheses).

	New products (N=79)	Improved products (N=50)
Flexibility[1]	2.76 (0.49)	2.60 (0.61)
Openness	2.71 (0.48)	2.72 (0.54)
Cooperation between departments	2.67 (0.47)	2.68 (0.51)
Supportive management style	2.61 (0.49)	2.56 (0.54)
Human focus	2.44 (0.59)	2.52 (0.61)
Focused on results	2.84 (0.37)	2.80 (0.45)
Creativity	2.77 (0.42)	2.76 (0.48)

[1] Scale of the variables is 1 to 3.

Table 2.5. Short- and long-term market performance for all products (N=129), improved products (N=50) and new products (N=79) (frequencies).

	Short-term market performance			Long-term market performance		
	Whole sample	New products	Improved products	Whole sample	New products	Improved products
High	38%	43%	31%	73%	74%	73%
Medium	49%	43%	59%	67%	62%	72%
Low	13%	14%	10%	25%	27%	20%
Total	100%	100%	100%	64%	62%	67%

on the market seven years after announcement. Products with a medium short-term market performance have a considerable market sustainability as 67% (of the 49%) are still on the market after seven years; while products with a high short-term market performance have the highest chance to be still on the market in the long term (73% of the 38%).

Similar conclusions can be drawn for the groups of new and improved products separately. There are only small differences: new products perform better in the short term and improved products perform better in the long term. Mann-Whitney analyses show that these differences are not significant.

Correlation analyses confirm that there is a significant positive relationship between short- and long-term market performance for the whole sample (See Table 2.6). This also applies to the two product groups: the relationship is highly significant for the group of new products (coefficient $=0.28$; $P=0.007$) and slightly significant for the group of improved products (coefficient $=0.22$; $P=0.067$).

The unexpected high success rate of both new and improved products in our sample compared to what is found in the literature might reflect the method we have used for the identification of new products. The products were selected by the editorial board of the professional journals. It can be assumed that they have chosen products which are worth being announced or mentioned as these are interesting new products with a good chance of being successfully marketed or are interesting improvements on existing products. However, the most likely explanation could be that the products in our sample are products that survived the critical 'first year slot', where one out of three newly launched products fails (Van Poppel, 1999).

Table 2.6. Correlations diagram for the control and independent variables (N=129).

	1	2	3	4	5	6	7
Control variables							
1. New product							
2. Product for consumer market	0.18 **						
Independent variables							
3. Technical feasibility	0.17 *	0.17 **					
4. Test production process	-0.02	0.31 ***	0.29 ***				
5. Marketing research	0.07	0.23 ***	0.38 ***	0.46 ***			
6. Test product concept	-0.08	0.09	0.33 ***	0.29 ***	0.25 ***		
7. Scale of preparatory and test activities	0.17 *	0.29 ***	0.56 ***	0.45 ***	0.51 ***	0.46 ***	

Spearman correlations (2-signed); significant correlations are indicated by *** for $P<0.01$, ** for $P<0.05$ and * for $P<0.10$.

2.4.2 Determinants for short- and long-term market performance

The correlation analyses between short-term market performance and the variables for upfront activities, organisational routines and company culture show that there are considerable differences between the innovation process aspects that explain the short-term market performance of new products versus that of improved products. Figure 2.2 presents the variables that have a significant relationship with short-term market performance.

For new products, our analyses show that ideas for new products coming from the board of directors of the company is a crucial factor for short-term market performance, which might reflect the importance of top level commitment (Kanters, 1984; Lester, 1998; Anderson, 2008). Performing a considerable number of preparatory and test activities – both technology-related and market-related – relates positively to short-term market performance. The involvement of the company's production, logistics and marketing functions also plays a significant positive role in the new product's short-term market performance.

The use of a product development plan is also very significantly correlated with the short-term market performance of new products. This also applies to the scale of preparatory and test activities: the more of these activities are being performed, the better the product's short-term market performance.

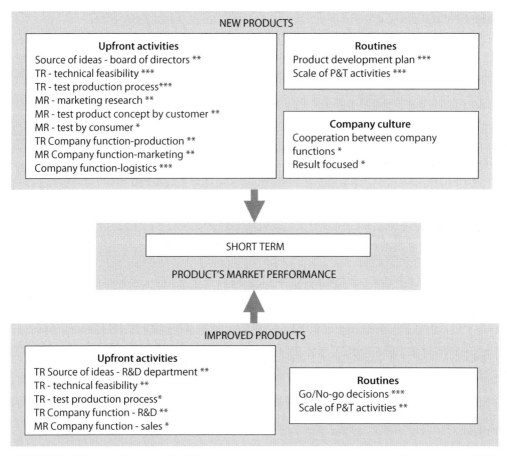

Figure 2.2. Innovation process variables significantly related to short-term market performance of new and improved products.
Spearman rank correlations (one-tailed): in the figure only significant relations are included; they are indicated by *** for P<0.01, ** for P<0.05 and * for P<0.10.
TR = technology-related; MR = market-related; P&T = preparation and test activities.

Although we found hardly any differences for culture characteristics between the two product groups (see Table 2.4), correlation analyses between each of the culture variables and short-term performance show that two cultural aspects – supporting cooperation and focussing on results – relate significantly positive to the new product's short-term market performance.

For improved products, the R&D department is a critical source of ideas for the product's market performance. Only two preparatory and test activities play a key role: both are technology-related (technical feasibility and testing production). The involvement of the company's R&D and sales functions each relates positively to short-term market performance of improved products. Also the scale of preparation and test activities is a success factor, but less significant than for new products. Improved products differ from new products in the role of go/no-go decisions; they are a crucial success factor for improved products. None of the company cultural characteristics seems to play a significant role in the short-term market performance of improved products.

Correlation analyses between long-term market performance and variables for technology-related and market-related sources of ideas, preparatory and test activities and company departments involved and for organisational routines and company culture characteristics again show significant differences between new and improved products (see Figure 2.3).

Interestingly, for new products, none of the upfront activities, routines or cultural characteristics relates significantly to long-term market performance. For improved products two market-related preparatory and test activities play a significant role in the product's long-

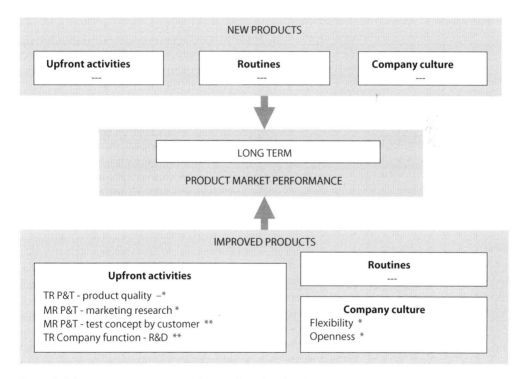

Figure 2.3. Innovation process variables significantly related to long-term market performance of new and improved products.
Chi-square coefficient: Phi for 2x2 contingency tables, Cramer's V for 2x(2+n). In the figure only significant relations are included and are indicated by ** for P<0.05 and * for P<0.10.
TR = Technology-related; MR = Market-related.

term success: marketing research and concept-testing by customers. The involvement of the company's R&D function also plays a significant role in the long-term market performance of improved products. For the long-term success the contribution of the sales function seems to be of less importance. Also none of the variables dealing with procedures play a role in the long-term market performance. Analysis of the role of cultural aspects shows that a flexible and open culture plays a slightly significant positive role in the long-term market performance of improved products.

We found a negative relationship with quality testing of improved products. We could explain this by suggesting that there might already have been some safety concerns at the time when the company developed the product. Since the late 1990s food and safety issues have become increasingly important and have attracted growing public attention, therefore these products might have been withdrawn from the market.

2.4.3. Importance of in-house technology- and market-related resources

In addition, we analysed the relative importance of technology- and market related resources against each other for the whole sample. For this we performed a regression analysis with as independent variables two technology- and two market-related resources that for both product groups were most significantly related with the product's market performance. Based on the outcome of the correlation analyses we selected technical feasibility and test of production process as they are both significantly related to the short-term market performance of new and improved products. The two most important market-related resources are marketing research and the testing of the product concept by potential customers as they are significantly important for short-term market performance of new products and for the long-term market performance of improved products. We also include the variable scale of preparatory and test activities, which proved to be significant for the short-term market performance of new and improved products.

Control variables are a dummy variable for product innovativeness (1 for new products; 0 for improved product) and a dummy variable for type of market (1 for consumer market; 0 for industrial and food service markets). A correlation analysis to test for multicollinearity among the independent and control variables is reported in Table 2.6. As the largest correlation is 0.56 and most of the remaining correlations are considerably lower, multicollinearity is not a significant problem in the analysis (Hair *et al.*, 1998). Furthermore, for the linear regression analysis (Model 1) all VIF values are checked (not included in the table); they are all between 1 and 2, except for the variable scale of preparatory and test activities which is 3.45. This is far below the upper threshold value of 10 (*ibid.*), which also indicates that multicollinearity problems are not encountered.

Table 2.7 presents the results of the linear regression analysis for short-term market performance (Model 1). Standardised Beta coefficients are displayed (ranging between 0 and 1) for the

Table 2.7. Regression analyses on in-house technology- and market-related resources with short- and long-term market performance (N=129).

	Model 1: Short-term market performance [1]	Model 2: Long-term market performance [2]
Control variables		
- New products	-0.01	0.12 (0.42)
- Products for consumer market	0.11	1.02 (0.45) **
Independent variables		
Technology-related resources		
- Technical feasibility	0.19 *	-0.55 (0.58)
- Test production process	0.10	0.70 (1.02)
Market-related resources		
- Marketing research	-0.13	-0.34 (0.74)
- Test product concept by customer	0.02	-1.19 (0.73) *
Number of preparatory and test activities	0.19	0.05 (0.13)
Constant	3.61 (0.53) ***	-0.19 (1.87)
R^2	0.14 ***	
Adj. R^2	0.10	
Pseudo R^2		0.15 **
Chi^2 - statistics		14.60
-2 Log likelihood		153.707
F	3.311	
Df	126	

Significant relations are indicated by *** for P<0.01, ** for P<0.05 and * for P<0.10.
[1] Standardised regression coefficients are displayed; for the constant term the beta value (B) and its standard error between parentheses are displayed.
[2] Beta values (B) and their standard errors between parentheses are displayed.

control variables and for the independent variables. Model 1 is significant and explains 14% of the variance in the short-term market performance. One variable is significant in the model explaining short-term market performance; this is the technology-related variable technical feasibility. This finding is in line with what we found in the previous section for the two product groups: that overall more technology-related than market-related resources are associated with and thus could explain short-term market performance. Here we find that a technology-related resource is more important than market-related resources for short-term market performance.

Model 2 in Table 2.7 shows the results of the binary regression analysis for long-term market performance. Beta values and standard errors are displayed. The model is significant and the value of the Nagelkerke pseudo R^2 which can be interpreted in a similar way to R^2 in linear regression analysis shows some predictive power. Surprisingly, it showed that when measuring the relative importance of testing of the product concept by potential customers against the other market-related and technology-related resources and the number of preparatory activities for the whole sample, it has a negative effect on long-term market performance. Apparently, testing of the product by potential customers is even counter-productive for long-term market performance, which is rather understandable given the large time span of seven years and the expected changes in consumer preferences during such a period.

The control variable Consumer market shows there is a positive association between products for the consumer market and long-term market performance; apparently products for the consumer market perform better in the long term than products for the other markets (industrial and food service).

2.5 Discussion and conclusions

The discussion in literature on the importance of involving technology- and market-related resources in the innovation process of new and improved products is very interesting as both represent important enablers of the innovation processes in the F&B industry. However, there is no agreement on the importance of each type of these resources for the two product groups. Balanchadran and Friar (1997) state that technology-related resources are the most important for new product development while Earle *et al.* (2001) argue that for new products, market-related resources are most important. A better understanding of the role of these resources in the new and improved product's market performance can be helpful in optimising innovation management and ultimately increasing the success rates of F&B products on the market.

The new products' short-term market performance is determined by more different and more significant resources than that of improved products; both technology- and market-related resources (activities and company functions involved) play a significant role. For the improved products we found that market-related resources only play a minor role in the products' short-term market performance: the involvement of the company's sales function will do. This is in accordance with Earle *et al.* (2001) who argue that as improved products were already on the market in an earlier version, the market is a *terra cognita* which has already been conquered. For that reason the involvement of market-related resources is not crucial for the product's market performance anymore. In addition, we found that for the improvements of the product to be made, the involvement of technology-related resources (activities and company functions) in the idea stage and in the preparatory and test stage of the innovation process proved to be crucial for the short-term market performance of improved products. These technology-related resources are involved in order to improve the product's characteristics which are a determinant of successful continuous innovation on which many studies agree (Van der

Panne, 2004). This finding seems not in accordance with Balanchadran and Friar (1997) who state that these resources are most important for new products and less for improved products.

The regression analysis confirmed this as it showed for the whole sample that technology-related resources – *in casu*: technical feasibility – are significantly more important for the product's short-term market performance than market-related resources. The importance of the R&D function for successful short- and long-term market performance of improved products is confirmed by Santamaría *et al*. (2009) who found that in-house R&D is shown to be of major significance for all types of innovation output in both low-, medium- and high-tech industries. We found that for more sustainable growth both technology- and market-related resources are needed: for the successful long-term market performance of improved products – these are the cash cows of a F&D company – not only the R&D-function needs to be involved, but also market-related activities (marketing research, testing by customers) turn out to be of significant relevance for the product's sustainable success on the market.

With respect to organisational routines it was found that a number of them have been of great importance for the short-term market performance of both new and improved products. For the development of new products the use of formal product development plans and for improved products the use of go/no-go decisions are crucial for the product's short-term market performance. For both product groups the more concise and extensive the upfront activities of their product development process, the better the short-term performance on the market. These results for new products are not in accordance with Earle *et al*. (2001) or Balanchadran and Friar (1997). Both stress the importance of organisational factors for improved products: standard procedures are suitable for improved products, new products require flexibility and creativity and less standardisation. Also, other authors found that formally organised structures discourage product champions (e.g. Howell and Higgens, 1990). However, most found them to be supportive as they enable the product development process and institutionalise and facilitate communication between the different company functions (Cooper and Kleinschmidt, 1987; Calantone *et al*., 1993; Lester, 1998). Our findings confirm this and also show internal consistency especially for new products where both the use of product development plans and the number of preparatory and test activities are highly significantly correlated with short-term market performance.

A most surprising result of our study was that for the successful long-term market performance of new products none of the innovation process variables (upfront activities, routines) proved to be significant. We conclude that – also due to the many major uncertainties that go along with the development of new products – other innovation trajectories and management systems can play a key role in the long-term market performance of these products. Unsurprising is the finding that a company culture that favours cooperation (especially between the technology-related and market-related company functions) and that is focused on results is a prerequisite for a new product's short-term market performance. However, what is striking is that for improved products to stay successful on the market in the longer term, the company culture that is characterised by flexibility and openness is of such significance while these cultural characteristics are often associated with new product development (see also Earle *et al.*, 2001). These cultural aspects deal with conditions that are closely connected with companies that want to stimulate creativity, capitalise on new chances, give room for individual initiatives and facilitate the trial-and-error character of the new product development process (Peters and Waterman, 1982). However, our study shows that these are also crucial for companies that want to perform a high level of continuous innovation, specifically the constant improvement of their core products.

3. Innovation process II: consumer versus industrial products

3.1 Introduction

This chapter presents a study on the differences in the innovation processes for consumer products and industrial products. The descriptive study investigates the role of technology-related and market-related factors in the innovation processes and the market performance of these products.

Studies on the innovativeness and success of products in the F&B industry on the market only focus on products for the consumer market (Traill and Grunert, 1997; Stewart-Knox *et al.*, 2003; Stewart and Martinez, 2002; Beckeman and Skjöldebrand, 2007; Van Trijp and Van Kleef, 2008) or cover all products – products for the consumer and the industrial market – but most do not discriminate between them (Galazzi and Venturini, 1996; Herrmann, 1997; Martinez and Briz, 2000; Menrad, 2004; Batterink *et al.*, 2006)[3]. Industrial products such as ingredients are often important components of consumer products (Earle, 1997). The consumer seldom notes the ingredients as they are usually not identified by the original manufacturer's brand or trademark on the consumer product label (Fuller, 2005). However, due to the growing market of ingredients for functional foods (Stein and Rodríguez-Cerezo, 2008) and – more specifically – the addition of healthy ingredients to consumer products (such as omega fatty acids or fibres) the name of the ingredient's brand is sometimes written on the product's label or used in advertising. This also applies to products in which unhealthy ingredients are replaced with ingredients that are regarded as less unhealthy (such as artificial sweeteners for sugar or fat substitutes for transfatty acids).

We argue that in order to fully understand the nature of product innovation in the F&B industry, one has to distinguish between product innovations for the consumer market and for the industrial market (see also Fuller, 2005). The product innovation processes and the specific factors that influence the speed and direction of these processes can be rather different for the two product groups; consequently innovation management may have to focus on different aspects.

The main research question addressed in this descriptive study is:

RQ2: *What are the differences in the involvement of technology-related and market-related resources in the innovation process of consumer versus industrial products and in their short- and long-term market performance?*

---

[3] Grunert *et al.* (2008) distinguishes between mass markets and business-to-business markets.

This study focuses on product innovations of companies in the Dutch F&B industry, which is the largest manufacturing sector in the Netherlands, representing 22.4% of the total turnover of all manufacturing industries (LEI, 2008; data for 2005). Within Europe the Dutch F&B is relatively very productive: the Dutch contribution to EU25's total turnover in this industry was 6.3%, while the Dutch F&B companies only constitute 1.8% of the total number of F&B companies in EU25 (INNOVA, 2008).

This chapter is organised as follows. Section 3.2 presents the theoretical framework and hypotheses. In Section 3.3 on data and methods we introduce the research design: an empirical study using data on 113 product introductions on the Dutch F&B market. Section 3.4 with the results of the study first presents the products in our sample (3.4.1). Subsequently, the role of technology- and market-related resources in the idea stage of the product innovation process (3.4.2) and the involvement of company functions and of external actors in the development stage of the product innovation process is presented (3.3.3). The section closes with the results dealing with the short- and long-term performance of both product types (3.4.4). Section 3.5 discusses the outcomes and draws conclusions.

3.2 Theoretical framework and hypotheses

Developments in science and technology and in the market are important factors that drive innovation processes. They push (technology) and pull (market) the innovation process – in which actors from inside and outside the company are involved – forward in a rather dynamic and interactive way. A company's innovative capabilities depend on its ability to manage and co-ordinate a variety of interdependent internal and external resources (Freeman and Soete, 1997). Although the open innovation concept is still rather new for the F&B industry (Sarkar and Costa, 2008; Smit et al., 2008), studies found that F&B firms rely more on external sources of innovation than the average for all industries (Archibugi et al., 1991). For the F&B industry the interactions of companies with their business partners in the supply chain, as well as with public research organisations, play a crucial role in achieving successful innovations. Suppliers of ingredients provide in-depth information on intrinsic aspects of ingredients such as flavours and fragrances; or antioxidants, fat and sugar replacers because of the health benefits (Joppen, 2004). They might even become more important for innovation than the suppliers of equipment and machinery that were traditionally the most important innovation sources in this industry (Christensen et al., 1996; Rama, 1996; Martinez and Briz, 2000). F&B companies have also developed networks with companies that provide them with market intelligence through which they keep track of their end-users and explore future consumers' trends, but they are also directly related to retailers and consumers (Grunert et al., 1997; Traill and Meulenberg, 2002). Knox et al. (2001) found that wide consultation with actors and the involvement of expertise beyond the company had a positive impact on the success of F&B products.

Although consumer products and industrial products are part of the same F&B production chain, we expect that the technology- and market-related resources will drive the consumer product innovation process in different ways and partly also by different actors as compared to the industrial product innovation process. The types of technologies and the technology-related actors that provide them can be rather different; see for instance Earle (1997) on the various technological innovation streams for different types of food products. Market-related actors differ also and the relationships with the market-related actors. For instance in the case of consumer products, the customer that buys them and the consumer that eats them (end-user) is usually one and the same, while in the ingredients industry several different customers i.e. F&B manufacturers, can buy one and the same ingredient for making a series of different products that reach a multitude of different end users (Fuller, 2005).

Our study takes a new approach as it focuses on the innovation processes and the market performance of a specific product, and not on that of a company. Resources can be actors (organisations or individuals) and the products of these actors that embody valuable knowledge (such as patents, publications, equipment, ingredients, brand positioning, market assessments). These actors can be actors from within the company (representing the different company functions) and from outside the company. Another novelty in our approach is that we investigated the product's market performance not only soon after market launch, but also after several years.

We have formulated two hypotheses: one that relates the two product groups to the two groups of resources driving innovation and one on the expected short- and long-term market performance of the two product groups. The development of ingredients for functional foods, but also other ingredients, such as those used in bakery mixes (enzymes, yeasts) is relatively knowledge-intensive. Research is needed in order to investigate the functionality of the specific ingredient and develop formulations for optimal activity (Enzing and Van der Giessen, 2005). Moreover, the processing of the agricultural inputs into raw materials for the F&B industry (sugar, milk, starch, flour, etc.) is also rather technology-intensive. For that reason it can be expected that in the development of F&B ingredients and raw materials technology-related factors are involved relatively more often compared to consumer products. We expect that industrial products will be more successful in the long term as a longer period is needed to earn back the returns on the investments that have been made for the development of these products. Consumer products are expected to have a relatively better performance in the short term: their life cycle is shorter as regular product launches are necessary in order to catch the consumer's eye and keep the 'brand' alive.

Hypothesis 2a (H2a): *The innovation process of consumer products will involve more market-related resources, whereas that of industrial products will involve more technology-related resources.*

Hypothesis 2b (H2b): *Consumer products will show a better short-term market performance, whereas industrial products will show a better long-term market performance.*

3.3 Data and methods

The first part of the methodology followed in the present study was to collect data on new consumer and industrial products of the F&B industry by a systematic review of the issues of 11 different Dutch food trade and professional journals (e.g. VMT, Eyes on Food, Food Management, Food Press, Bakkerswereld) that were published in the second half of 1998. The basic element of the methodology is that innovations are identified by sampling the editorially controlled 'new product announcements' sections of these journals. The advantage of this method is that it has been the decision of the journal editors – an independent, qualified panel – to include them and not that of a stake-holding actor such as the commercialising company. Information was gathered from the journals about the product's name and the name of the company. The second step was to gather additional data by conducting phone interviews with managers of the companies who had been directly involved in the product development process. These data – on the sources of ideas for the product, the company functions and external actors that had been involved in the product development process and the product's performance on the market – were collected in the period March to September 2000. A structured questionnaire was used. Additional data on the long-term market performance of the products were collected in December 2005, seven years after product announcement. The Mann-Whitney test (2-Independent samples test) which is suitable for non-parametric data analysis was applied to test between two groups (consumer and industrial products).

3.4 Empirical results

3.4.1 Products in our sample

The sample of products for which complete datasets could be collected included 76 consumer products and 37 industrial products. The 113 products represent seven subsectors within the F&B industry: fruit and vegetable products; oils and fats; dairy products; grain mill products, starch and starch products; bakery and farinaceous products (macaroni, noodles, pasta, etc.); beverages and finally a group of 'other food products' which included ready-to-eat meals (including breakfast products) and snacks and products not included in the other food products categories. Products for the industrial market are in each of these subsectors, except for in the oils and fats subsector. Most are in the two subsectors 'bakery and farinaceous products' and 'other food products'. The group of industrial products includes both food ingredients and ready-to-make mixes and pre-mixes for (industrial) bakeries. See Box 3.1 for examples of consumer products and industrial products in our sample.

Product innovativeness

We measured the innovativeness of the products first of all by asking the companies: is the product new or is it an improved or renewed version of a product that was already on the market? About one-third of the products for the consumer market (32%) were perceived as

Box 3.1. Examples of consumer and industrial products included in the study.

Consumer products

    Grolsch 2.5

        Product information: A light lager (alcohol percentage of 2.5)

        Company: Koninklijke Grolsch N.V., Enschede

        (Food Press, 1998, no. 39, p. 4 and no. 52, p.4.)

    Fruitontbijt (Fruit breakfast)

        Product information: A ready-to-use breakfast product made from pure juice and fibres from whole grains.

        Company: Hero, Breda

        (Food Press, 1998, no. 36, p. 4.)

    Milner Gerijpt (Milner Ripened)

        Product information: A cheese with 30% less salt and 40% less fat. Due to a special recipe and natural maturation, the taste is similar to cheese without less salt and fat. The product was nominated for the Goede Voeding Prijs (Good Food Price) 1998.

        Company: Campina Melkunie (Divisie Kaas), Tilburg

        (VMT, 1998, no, 22, p. 8 and Eyes on Food, 1998, no. 5, p.55.)

Industrial products

    Paselli Easy Gel

        Product information: Modified potato starch product: the gel – which can be prepared cold without other gel supporting aids – can be cut within 30 minutes. Applications in pastry cream, fruit fillings and other instant products.

        Company: Avebe, Foxhol

        (Food Management, 1998, no. 12, p. 27.)

    Springline

        Product information: Quality flour with high content of protein and low dose of vitamin C for luxury bread products for special occasions.

        Company: Koopmans Meelfabrieken, Leeuwarden

        (Bakkerswereld, 1998)

---

new products, against more than half (54%) of the industrial products; the rest are improved products. The Mann-Whitney test showed that the two product groups differ significantly ($P=0.02$) for this newness variable. Next to this more subjective measure of product innovativeness we used a number of objective measures: newness of product attributes, newness of knowledge used to develop the product and newness of the production process. The respondents could indicate for which (combination of) attributes the product was new or improved: raw materials, ingredients, processing, recipe, shelf-life conditions, readiness to use/ eat, packaging and/or nutritional value. The values for the indicator 'New product attributes' are the sum of the positive scores on each of the attributes: the more new or improved product attributes, the higher its innovativeness. For the newness of knowledge that was used for the development of the product, the respondents were asked to choose between: has new scientific-technological or other knowledge been used that was developed by the company itself, by others, or has existing and earlier applied knowledge been used? For the newness of the production process the companies were asked: was a totally new production process developed; or has the existing process been adjusted (significant or small) or were there no major changes?

Table 3.1 shows that consumer products have a slightly higher innovativeness level for all three newness indicators; only for the product attributes indicator did the two product groups differ significantly from each other. This is also what we expected: the complexity of consumer product innovations is higher than that of products for the industrial market as – next to new product and process aspects which are comparable – consumer products can also include new attributes such as those that are related to storage conditions in the shops or its use by the consumer.

3.4.2 Idea stage

For investigating the role of technology-related and market-related factors in the product innovation process we focused on the first two stages of the product innovation process: the idea stage and the product development stage. Other stages such as the production process and market launch are not included. For the idea stage the use of internal and external sources of ideas for innovation was investigated: both are the 'feedstock' of the product's innovation process (Cooper, 1983).

Table 3.1. Product innovativeness: attributes, knowledge and process (means, standard deviation between parenthesis).

	Consumer products (N=76)	Industrial products (N=37)	DbG[1]
Newness of product attributes [2]	3.96 (1.06)	3.41 (0.80)	**
Newness of knowledge used [3]	1.63 (0.86)	1.61 (0.84)	
Newness of production process [4]	1.84 (1.04)	1.81 (1.02)	

[1] DbG - Difference between Groups. Significant difference between the two product groups (using Mann-Whitney test; excluding cases test-by-test; exact sig 1-tailed) is indicated by ** for: P<0.05.
[2] Indicator values range from 0 to 8.
[3] 3-point Likert scale.
[4] 4-point Likert scale.

Technology-related sources of innovation

For both types of products, in the idea-stage the company's R&D department is by far the most important technology-related source of innovation (see Table 3.2). From the five external technology-related sources of innovation, the suppliers of raw materials and ingredients are equally important for consumer and industrial products: in 16% and 17% of the cases they are mentioned as source of innovation. Suppliers of machinery and equipment play a marginal role in this phase and only for consumer products (5%). Interestingly, research institutions such as universities or public research institutes have also seldom been used as a source of innovation. The two technology-related external sources that have been mentioned relatively more frequently in the idea stage of industrial products include professional literature and patent literature. The two groups differ significantly from each other for having checked patent literature: this is done more often in the case of new products.

Market-related sources of innovation

The company's marketing/sales department is the most frequently mentioned internal market-related source of innovation. It was used in fairly equal measure by both product types: 92% for consumer products, 89% for industrial products (see Table 3.2). However,

Table 3.2. Sources of innovation for consumer and industrial products in the idea stage (frequencies).

	Consumer products (N=76)	Industrial products (N=37)	DbG[1]
Technology-related			
Company's R&D department	75%	70%	
Suppliers of machinery & equipment	5%	0%	
Suppliers of ingredients & raw materials	17%	16%	
Research organisations	1%	0%	
Professional literature	16%	22%	
Patent literature	1%	11%	**
Market-related			
Company's marketing/sales department	92%	89%	
Customers/consumers	21%	54%	***
Competitors	18%	30%	*
Fairs and exhibitions	13%	22%	

[1] DbG - Difference between Groups. Significant difference between the two product groups (using Mann-Whitney test; excluding cases test-by-test; exact sig 1-tailed) is indicated by *** for: P<0.01; ** for: P<0.05 and * for P<0.10.

the most interesting outcome deals with customers and/or consumers. Customers in our definition are companies that buy F&B products from other companies upstream in the food chain. Both producers of industrial products and of consumer products can be customers. Retailers and food service providers are also customers. Consumers in our definition are the customers of the producers of consumer products that buy the food products from retailers or food service providers for their own consumption. Industrial product companies have used customers relatively more often as source of innovation than consumer product companies: the two groups of products differ significantly in this respect. Also the other external market-related sources of innovation – competitors and fairs & exhibitions – are used relatively more often in the case of industrial products.

Overall, we can conclude that in the idea stage of the product innovation process the internal sources of innovations – R&D department and the marketing/sales department – were the most important sources for both product types and were used in equal measure. There are significant differences in the use of external sources of innovation: industrial product innovation processes used both technology-related (patents) and market-related (customers, competitors) sources of innovation relatively more often.

3.4.3 Development stage

We investigated the role of technology-related and market-related resources in the development phase of the consumer and industrial product innovation process first of all by measuring the involvement of specific company functions in this process and the use of cross functional teams. Apart from the R&D function, we also included the purchasing function, the quality and safety function and the production function as technology-related. Purchasing was included as innovative products that have been developed outside the company can be incorporated by the acquisition of ingredients, equipment or machinery. These products are an important interface through which F&B companies communicate with innovators in other industries and apply scientific advances that have been developed in these industries (Christensen *et al.*, 1996; Rama, 1996).

The food quality and safety function includes the use of many R&D-intensive methods, such as those dealing with the monitoring of potential hazardous microorganisms (Fuller, 2005). The production function was included as the product can demand a new or an adjusted production process (in fact: in about 50% of the cases the production process was changed). Market-related company functions included the marketing and the sales functions, as market demands and developments can be fed into the innovation process through these functions. We included the use of cross-functional teams as the coordination between the R&D function and the marketing function in such teams is an important factor for market performance of food products (Dahan and Hauser, 2001; Rudder *et al.*, 2001).

Differences in complexity of the product's internal innovation network

First of all our results showed that the development of consumer products took place significantly more often in cross-functional teams: in 88% of the consumer product cases versus 43% of industrial product cases.

Secondly, we found that in the development stage – as in the idea stage – the company's R&D function was deeply involved and also rather similar for both product groups (see Table 3.3). The other technology-related company functions (purchase, quality and safety, production) were more involved in the consumer as compared to the industrial product development process. The company's marketing function played a relatively less important role in the development stage as compared to the idea stage. However, the level of involvement of the marketing function differed significantly between the two groups: they were involved in about half of the cases for consumer products and in one third of the cases for the industrial products.

Overall, more different company functions were involved in the development of consumer products than industrial products. This is in accordance with what we found for the innovativeness of consumer products in terms of product attributes. Again the more complex character of consumer products is illustrated, now by the larger variety in different company functions that are involved in its development.

Table 3.3. Involvement of company functions and use of cross-functional teams for consumer and industrial products in the development stage (frequencies).

	Consumer products (N=76)	Industrial products (N=37)	DbG[1]
Technology-related company functions			
R&D	88%	87%	
Purchase	25%	14%	
Quality	23%	14%	
Production	30%	19%	
Market-related company functions			
Marketing department	49%	32%	*
Sales department	32%	35%	
Cross-functional teams	88%	43%	***

[1] DbG - Difference between Groups. Significant difference between the two product groups (using Mann-Whitney test; excluding cases test-by-test; exact sig 1-tailed) is indicated by *** for: P<0.01 and * for P<0.10.

We also investigated the differences in contribution (in terms of man-years spent) of the company's R&D-function and of the other company functions to the product development process. It showed that the contribution of both is significantly higher for the group of consumer products, which is in accordance with the finding that more functions are involved. We checked for the size effect (in our sample do we have more large consumer products producers and more small and medium-sized producers of industrial products, which might explain the difference), but found that there is a rather even distribution of the product types across SMEs and large firms. However, the data sets on input of the R&D function and of the other company functions are not complete; for that reason we hesitate to draw firm conclusions.

Innovation networks

We investigated the cooperation of F&B companies with other companies (suppliers, customers) and with public research institutions (universities, research institutes, polytechnics) in the product development process. These other companies and research institutions can be involved in three different roles: as partner, as outsourcer for specific activities and as seller from which specific products or services are purchased.

Our study shows that the networks of consumer products differ considerably from that of industrial products. The technology-related part of the consumer product's network is more open in terms of number and variety of partners, outsourcers and sellers that are involved, compared to that of the industrial product's network (see Table 3.4). The technology-related part of the industrial product's network only includes more research institutions to which activities are outsourced. Specific data on the research institutions to which the industrial product development process has been outsourced (not presented in the table) shows that this was mostly to research institutes and polytechnics. Universities are equally (but at a low level) represented in the networks of both product groups.

Companies producing consumer products cooperate significantly more often with producers of raw materials, ingredients, machinery or equipment as partners in the product development process than companies producing industrial products. Outsourcing to companies that provide recipe advice and training and to companies that sell machinery and equipment also happens relatively more often for consumer products than for industrial products, but these differences are not statistically significant.

Also, the market-related part of the product-related networks differs considerably between the two product groups. Customers of industrial products are significantly more often involved as partners in industrial product development processes as compared to the involvement of the customers of consumer products in the development of these products. The networks also differ significantly with respect to the outsourcing of marketing-related activities: this was in 57% of the consumer products' cases, and only in 8% of the industrial products' cases.

Table 3.4. Involvement of companies and public research institutions in the development of consumer and industrial products in the development stage (frequencies).

	Consumer products (N=76)	Industrial products (N=37)	DbG[1]
Technology-related			
Research organisation as partner	9%	5%	
Research organisation as outsourcer	9%	14%	
Producers of raw materials, ingredients, machinery, equipment as partner	46%	16%	***
Companies providing recipe advice as outsourcer	7%	0%	
Machinery/equipment producing companies as seller	31%	22%	
Companies providing training as seller	15%	8%	
Market-related			
Customers as partner	1%	35%	***
Consultancy company (including marketing) as partner	7%	3%	
Companies providing marketing-related activities as outsourcer	57%	8%	***

[1] DbG - Difference between Groups. Significant difference between the two product groups (using Mann-Whitney test; excluding cases test-by-test; exact sig 1-tailed) is indicated by *** for: $P<0.01$.

We also asked the companies what type of activities they had performed together with their partners. In 20 of the 35 cases where producers of raw materials, ingredients, machinery or equipment were involved as partners in the consumer product development process, they mainly worked together on ingredients (taste, texture, thickening, coating, quality, choice, recipe). Packaging (flasks, labels) was in five cases the object of cooperation. Other objects of cooperation included: process technology, production process, equipment and technical feasibility. In the only case where they had a customer as partner, this was for preparing the market introduction.

Furthermore, in one case of the industrial products group, market implementation was the cooperative activity with a customer. In all other industrial product cases where customers were involved (in total 35% of the 37 cases), this was for activities such as developing the basic product requirements and testing the product. Suppliers to producers of industrial products were involved as partners (six cases) mostly in the development and testing of ingredients and raw materials. Cooperation with research institutions normally involved development and specific testing of the products.

3.4.4 Short- and long-term market performance

The short-term market performance was measured by a combination of two indicators: impact of the product on the company's market share and on the company's turnover. The value of the indicator 'Short-term market performance' is the average of the scores on these two indicators: low performance stands for no change, medium stands for small increase and high for large increase. We found (see Table 3.5) that nearly two thirds of all 113 products that were announced in 1998 were still on the market seven years later (long-term market performance). Interestingly, 20% of the products with a low short-term market performance were still on the market seven years later. Products with a medium short-term market performance have considerably higher market sustainability: 66% of these products were still on the market after seven years. However, as expected, products with a high short-term market performance have the highest chance of still being on the market in the long term (73%).

When analysing the two product groups separately we found that industrial products perform better than consumer products in the long term for all three short-term performance levels. In the short term they perform better for the medium and low performance groups. Consumer products only perform better for the high performance short-term group. Mann-Whitney

Table 3.5. Short- and long-term market performance of all products and consumer products and industrial products separately (frequencies).

	Short-term market performance [1]			Long-term market performance [2]		
	Whole sample	Consumer products	Industrial products	Whole sample	Consumer products	Industrial products
High	40%	46%	28%	73%	71%	80%
Medium	47%	43%	55%	66%	52%	90%
Low	13%	11%	17%	21%	7%	33%
Total	100%	100%	100%	63%	57%	76%

Whole sample: N=113; Consumer products: N=76; Industrial products: N=37.
[1] Short-term market performance: change in financial and market impact of the product (using a 3-point Likert scale: no change, small increase, large increase) in company turnover and market share respectively due to product introduction on the market, measured one and a half years after product announcement.
[2] Long-term market performance: percentage of products still on the market seven years after product announcement (% of the short-term market performance cases).

analyses (exact, 1-tailed) showed that the two product groups differ significantly from each other for both short-term performance (P=0.040) and long-term performance (P=0.037).

3.5 Discussion and conclusions

This study explores the differences in the product innovation process and market performance of consumer versus industrial products. Overall we found considerable differences. Consumer products' innovation processes used more different sources of ideas (see Table 3.2) and the external network also involved in the product development process of these products was more open as it included more different partners (see Table 3.4), as compared to industrial products.

When considering the composition of the internal and external product-related innovation networks and more specifically the involvement of technology- and market-related resources, the results of our study only provide partial evidence that supports the first part of our first hypothesis (H2a) which states that the innovation process of consumer products includes relatively more market-related resources than that of industrial products. In the idea stage the company's marketing and sales function was only slightly and not significantly more often used as source of innovation. The most supportive results were found for the development stage: here significantly more often the company's marketing function was involved and also marketing-related activities were more often outsourced. However, we also found evidence to support the opposite. Firstly, customers, but also competitors, were used significantly more often as source of ideas for industrial products. Secondly, producers of industrial products worked together with customers/consumers as partner in the development stage significantly more often. Our conclusion is that the first hypothesis cannot be confirmed.

The second part of the first hypothesis says that the innovation process of industrial products includes technology-related factors relatively more often than that of consumer products. Our results showed that in the idea stage patent literature was used significantly more often as source of ideas for innovation. Professional literature was also used more often for industrial products, but not significantly. In the development stage research institutions were involved more often as partner in the industrial product's development process but the difference with consumer products is not significant. However, our results also showed that technology-related aspects are important in consumer products innovation processes. First of all, in the idea stage the R&D function as source of innovation was used slightly more often for consumer products than for industrial products. In the development stage all four technology-related company functions were more involved in the development of consumer products. Also, the significantly more frequent involvement of producers of ingredients and raw materials as partners in the consumer product development process, confirms the importance of technology-related factors of consumer product development as the R&D-efforts of these suppliers are 'embodied' in this cooperation and in the products (Dosi *et al.*, 1990; Rama, 1996). Overall, we can conclude that the second part of the first hypothesis cannot be confirmed either, since we found strong evidence that technology-related aspects also play an important role in the consumer product innovation process.

The second hypothesis (H2b) states that industrial products have a better market performance in the long term and consumer products in the short term. This hypothesis could be confirmed for the industrial products that showed an overall better performance in the long term. However, in the short term they also performed better than consumer products, except for the group of products with a high level of short-term performance. Apparently, consumer products not only stay for a shorter time on the market, but also contribute less to a company's turnover and market share, compared to industrial products. One explanation for this could be that one ingredient can be part of a large diversity of consumer products: their life cycle consists of several parallel streams of which some can also be much longer than others.

Our findings that customers are significantly more often involved in the innovation process of industrial products both as source of innovation in the idea stage and as partner in the development stage or, better still, the non-active role of customers/consumers in both stages of the innovation process of consumer products, are most surprising. Many sources – most focussing on consumer products – also stress the importance of the consumer-driven or user-oriented character of innovations in the F&B industry (including Galazzi and Venturini, 1996; Christensen et al., 1996; Grunert et al., 1996, 2008). We expected that the closer to the market, i.e. the consumer, the more often consumers would have directly or indirectly (through retailers) played a role in the product innovation process. This is not what we found. Grunert et al. (2008) explain this by arguing that the food products market is a mass market and it is impossible to innovate in interaction with all users. This market is at 'arm's length' and only in indirect ways information about users' preferences is collected, mostly using quantitative methods to characterise the different sub-populations of users. This could also explain why we found that only the company's marketing function played a significant role in this respect.

Market intelligence in this sector is essential for creating long-term competitive advantage, but Costa (2003) found that it is still relatively poorly developed. Other studies showed that most companies rely heavily upon the retailer to obtain market information about their end-users (Hoban, 1998; Parr et al., 2001; Knox et al., 2001; Stewart-Knox et al., 2003). However, on the basis of our study we could also conclude that the strong relationship of producers of consumer products with their ingredient suppliers (see Table 3.4) might also represent an important channel through which market knowledge is being transferred. Joppen (2004: 31) supports this as he found, on the basis of a survey of heads of product development in the F&B industry, that 60% of the respondents – especially from larger companies – acknowledge that suppliers have filled an important information gap by providing information on the consumer market. This information gap was perceived by 15% of the consumer product companies as the main obstacle in successful product development, compared to 4% of the suppliers. Earle et al. (2001) argue that food ingredient suppliers have gone a step further than just having a good relationship with the food manufacturer. They develop and produce the ingredient, but also design the manufacturing process for the producer of the consumer products; they hand over the complete package to the manufacturer. Earle *et al.* (2001: 114) suggest that the reason for this may be the greater knowledge of product development in the food ingredient companies. We would add: and also greater knowledge of the market for which these products are developed (and adding more value to the product for which they can ask a higher price). So market intelligence could be gathered by consumer product companies by a number of different channels; not only through interactions (of the company's marketing and sales function) with retailers or consumers directly but also indirectly, for instance through their ingredient suppliers.

Studies on the management of innovation processes across industries showed that product development in cross-functional teams is an important factor for the product's success (see for instance Cooper, 1983; Madique and Zirger, 1984; Link, 1987). This was confirmed for food products – and more specifically consumer food products – in many studies (see for instance: Hoban, 1998; Knox *et al.*, 2001; Stewart-Knox and Mitchell, 2003; Stewart-Knox *et al.*, 2003). The results of our study confirm this: the development of consumer products took place significantly more often in cross-functional teams as compared to that of industrial products.

Our findings that technology-related resources play an important role in the development of both consumer and industrial products, refers to the R&D-intensity of the F&B industry. Although this industry might have relatively low R&D-intensity according to formal statistics (Hollanders and Arundel, 2005), in both consumer and industrial product innovations the technology-related aspects were very important. This could be explained partly by the innovativeness of the Dutch F&B industry. The R&D intensity of the Dutch F&B industry was relatively high compared to that of other European countries over the period 1995-2002 and in 2002 even the highest: 0.6% against an European average of 0.24% (CIAA, 2006). Also, the results of a benchmark study that measured the innovative performance of a number of industrial sectors through a combination of indicators clearly showed that for the F&B sector the Netherlands takes the lead position: it has both a high level of activity (highest of all EU25 countries) and a high (second position) growth rate (INNOVA, 2008). However, the importance of technology for innovation processes in the F&B industry is a trend that has also been mentioned by others (Rama, 1998; Wilkinson, 1998; Traill and Meulenberg, 2002). F&B companies are increasingly confronted with competition from private label products and – in order to become more competitive – they choose to become more innovative, preferable in products and processes that have proprietary elements that can be protected. This is one of the reasons why the introduction of relatively high value added products such as functional foods have become increasingly important for the F&B companies; they raise entry barriers. Our findings, together with the argumentation that R&D may be underestimated in this industry because of a lower degree of functional specialisation in small and medium-sized firms (Kleinknecht *et al.*, 1991), and the growing market of knowledge intensive health ingredients might indicate the increasingly technology-based character of product innovations in the F&B industry. The results of our study underline the need for revising the market-dominated view of innovation processes in the F&B industry, and ask for a more detailed investigation of how and through which actors technology-related and market-related resources are influencing industrial and consumer product innovations in the F&B industry.

To date, this is one of the first studies focusing on the involvement of technology- and market-related resources in the innovation process of consumer versus industrial products of the F&B industry. Although some studies draw attention to the different markets of the F&B industry (Grunert *et al.*, 2008) and advice on the management of the product development process of consumer and industrial products (Earle *et al.*, 2001), this is the first study to analyse the two product groups separately. Moreover, scarce empirical evidence is available on the specific composition of the innovation networks of these product groups. As such, we found importance differences in the involvement of technology- and market-related resources in the internal and external networks of consumer versus industrial products. Further research may investigate more specifically how the internal and external networks are related and look for ways to improve cooperation and open innovation.

4. Innovation strategy: prospector versus analyser or defender strategy

4.1 Introduction

This chapter presents the results of the study on the impact of the company's product innovation strategy on the product's short- and long-term market performance.[4]

Companies increasingly operate in a competitive environment in which changes in competitors' strategy, new technological opportunities, changes in market trends or new governmental regulations can make a difference between success or failure. Those companies that stay alert to what is happening in their environment and that closely assess potential future developments are able to keep ahead of their competitors. As Jack Welch, CEO of General Electric, stated very clearly: 'If you do not control your own destiny, somebody else will do it' (De Vaan *et al.*, 1998). Empirical investigations have shown that an articulated strategy with a clear mission is an important factor for success. A strategy provides guidelines for dealing with strategic questions such as which new products to develop, for which markets, which expertise to develop within the company and which expertise to acquire from outside (Lester, 1998). Cottam *et al.* (2001) found that in order to maximise the benefits of previous innovation processes, companies must give these processes a strategic direction.

In this study we focus on the innovation strategies of the F&B industry. Companies in this industry increasingly have to compete on the basis of new and more advanced products (Menrad, 2004). We argue that product innovation strategy is becoming an element of growing importance in the F&B companies' overall strategy. Early studies on strategies of F&B companies deal with the extent of geographical market coverage and differentiation, entry barriers and bargaining power (McGee and Segal-Horn, 1992; Hyvönen, 1993). More recent studies also address innovation as a central aspect of F&B companies' strategies (Gilpin and Traill, 1999; Martinez and Briz, 2000; Traill and Meulenberg, 2002; Avermaete *et al.*, 2004; Batterink *et al.*, 2006). The present study goes a step further. By following an outside-in approach we have investigated how the performance of product innovation strategies of Dutch F&B companies can be measured from their outputs and how this relates to the innovativeness of the product. Unique aspects of the study are that it is focused on the product level and that it not only investigates the impact of product innovation strategy on the product's market performance in the short term, but also in the long term.

---

[4] This chapter is based on: The impact of the product innovation strategy on the product's short and long-term market performance: evidence from the Dutch food and beverages industry, Enzing, C.M., Batterink, M.H., Janszen, F.H.A. and S.W.F. Omta. A revised version of this article, based on comments of two anonymous referees has been resubmitted to the International Journal of Technology Management.

The main research question that is addressed in the study is:

RQ3: *What is the impact of the product innovation strategy on the product's short- and long-term market performance?*

We have structured the chapter as follows. Section 4.2 presents the theoretical framework and the hypothesis to be tested. Section 4.3 describes the methods of data collection, the variables that are used and data analyses. Section 4.4 presents the main empirical results. The last section discusses the main findings and the conclusions that can be drawn from the results as an answer to the central research question presented above.

4.2 Theoretical framework and hypothesis

4.2.1 Innovation, strategy and performance

Innovation

Innovation, as Schumpeter (1939) defined it, is 'any doing things differently in the realm of economic life'. So essentially, innovation is about change: change in the products or services of a company and change in the way the company produces them (Tidd *et al.*, 2005). There are degrees of change; from only minor incremental improvements, adaptations or refinements of existing products and processes to very radical changes leading to totally new products or production processes.

Innovations can contribute in many ways to a company's performance. Souder and Sherman (1994) showed for product innovations that there is a strong correlation between new products and market performance as they are a means to enter and retain new markets and increase profitability in these markets. Luchs (1990) found that differentiated products on both quality and other features were twice as successful in terms of return on investment.

Innovation strategy

A company's innovation strategy forms an essential element of the overall business strategy. Basically corporate strategy is about finding the perfect fit between the uncertain and continuous changing external (industry/competitive) environment in which a company operates and how the company's operations should be optimised in order to realise its mission. The corporate strategy sets the company's overall direction according to its mission and deals with the business the company wants to be in. The corporate innovation strategy deals with how this can be accomplished by the development of new products and new processes. Product innovation strategies specifically focus on which products the company wants to develop for which markets and through which technologies.

Miles and Snow (1978) have developed a theoretical framework that deals with the alternative ways organisations define their product-market domains and construct mechanisms to pursue these strategies. Their framework deals with the overall company strategy and has an important innovation component as it explicitly addresses the innovativeness of products and the role of technology and market in the development process of products. Based on Weber's (1947) structural contingency theory, Miles and Snow belong to the stream of research which classifies firms into schemes based on their environmental adaptation patterns (Ketchen *et al.*, 1997). According to Miles and Snow (1978) adaptation takes place in cycles each time the company or business unit has solved three basic problems: (1) the entrepreneurial problem (i.e. what business are we in?), (2) the engineering problem (i.e. which technologies can we use?) and (3) the administrative problem (i.e. how to facilitate and coordinate diverse operations?). Miles and Snow (1978) have defined three strategic types of organisation: prospectors, analysers and defenders. Each type has its own unique strategy for relating to its chosen markets and a particular configuration of technology, structure and process that is consistent with its market strategy. Prospectors typically foster growth by developing new products and exploiting new market opportunities. They invest in scanning the environment for potential opportunities. They are frequently the creators of change in their respective industries. Defenders strive to prevent competitors from entering their part of the market; they are hardly involved in new product development or in entering new markets. They choose to grow mainly by producing and distributing their goods as efficiently as possible. They have a stable set of products and customers; their product development is closely related to current goods. Through continuous technological improvements they produce as efficiently as possible. The third type – the analyser – combines the strengths of both the prospector and the defender. An analyser moves to new products or new markets only after their viability has been demonstrated. While the majority of the analyser's revenues are generated by a stable set of products and markets; periodically they extend their domain by imitating successful products that have been developed by prospectors. Miles *et al.* (1978) state that these three types of strategies show a stable and consistent pattern of response to the changing environments in which they operate. A fourth type of organisation – the reactor – applies the 'residual' strategy, arising when one of the other three strategies is improperly pursued.

Innovation strategy and performance

There is extensive literature on the strategy-performance relationship. Following the seminal work of Rumelt (1974) the main focus has been on how different types of market differentiation strategies have an impact on the company's profitability. He found that diversification in itself was not a profitable strategy, but that firms that pursue a diversification strategy that is related to their main activity, are more successful than companies that follow a diversification strategy that is unrelated to their main activities (*ibid.*). Recent studies found that companies that succeed in implementing a strategy that strikes the right balance between explorative innovation and effective exploitation of innovation can realise optimal value creation (Jansen *et al.*, 2006).

Miles and Snow (1978) postulated that although the three strategy types are different in their adaptive pattern – the defender protective, the prospector aggressive and the analyser analytical and prudent – each of these types is consistent in its behaviour across the cycle and is expected to perform well. However, empirical studies on strategy success do not come to uniform conclusions. McDaniel and Kolari (1987), Snow and Hrebinaik (1980) and Conant *et al.* (1990) found that defenders and prospectors perform equally well. Other studies have found that prospectors outperform defenders: Moore (2005) for the retail industry and Hambrick (1983) for innovative industries. Kearns (2005) found that in electronic commerce prospectors and analysers perform better than reactors.

4.2.2 Innovation, strategy and performance in the F&B industry

Until the early 1990's the F&B industry was classified as a scale-intensive process industry, according to the taxonomy introduced by Pavitt (1984). Innovation mostly dealt with the production process: companies in this large-scale production category produced a relatively high proportion of their own process technology. With respect to product innovations, their incremental character was considered as one of the key features of the F&B industry (Christensen *et al.*, 1996; Galizzi and Venturini, 1996). Studies on the German, Spanish, European and USA F&B markets showed that only a small portion of the product releases are truly innovative (Gallo, 1995; Rudolph, 1995; Connor and Schiek, 1996; ECR Europe, 1999; Martinez and Briz, 2000; Menrad, 2004). The incremental character of product innovations in the F&B industry was considered to be very inherent to the F&B product itself as consumers reveal a specific form of risk aversion in their choices: new food products have to be rather similar to familiar products (Galizzi and Venturini, 1996). Arguments for this position are not only found in the resistance to the use of specific new technologies such as gene technologies (Beckeman and Skjöldebrand, 2007) or radiated foods. More nutritional, biopsychological and cultural constraints also impose continuity on the demand side; taste and taste aversion such as the preference for sweetness and the abhorrence of bitterness, have an innate bio-physiological background (Rozin, 1987).

However, since the mid-90's a number of specific developments have had a profound impact on the character of innovation in the F&B industry. Two trends in particular have led to a deceleration and qualitative change in food demand: the increase in disposable income and the decrease in the population growth in developed countries. In addition, globalisation of large-scale food production and distribution – facilitated by scientific and technological developments – made a large diversified food supply almost permanently available in the western world (Costa, 2003). This implied that the importance of price and availability as determinants of food purchase decreased and the relative importance of consumer choice increased (Van Trijp and Steenkamp, 1998). Nowadays changing consumer preferences have become one of the main drivers for the expansion of the F&B industry (Galizzi and Venturini, 1996; Grunert *et al.*, 1997; Alfranca *et al.*, 2001). Food consumption patterns have changed more than ever over the last period. Due to changes in life styles, new household equipment

(Oldenziel, 2001), new beliefs (some also scientifically proven) about the preventive health aspects of nutrition and the process of increased internationalisation leading to more contact with foreign culinary traditional food (Grigg, 1995), diets have undergone dramatic changes. Consumers ask for ready-to-eat meals, healthy food (not only as a prevention strategy, but also to combat obesity), snacks and more exotic food and there is also a growing demand when it comes to food quality and food safety (Steenkamp and Van Trijp, 1996; Earle, 1997; Joppen, 2004; Menrad, 2004). Moreover, F&B companies are increasingly confronted with competition from private label products.

These trends have initiated firms in the F&B industry to reformulate their competitive strategy towards more demand-driven product diversification strategies and become more pro-active. One of the best strategies is to be innovative, preferably in products and processes that have proprietary elements that can be protected. This is one of the reasons why the introduction of relatively high value added products has become important for F&B companies (Menrad, 2004): this has enabled them to raise entry barriers. A study that analysed the innovation strategies of large European companies showed that in the F&B industry product innovations were considered more important than process innovations (Arundel *et al.*, 1995). This was confirmed by a company survey: about one third of the F&B companies in Germany that had innovation activities only had product innovations, 40% spent more than 50% of their innovation budget on product innovations, while about one quarter had spent more than 75% on process innovations (Menrad, 2004). In the Spanish F&B industry product innovation objectives were also found to be more important than process innovation objectives (Martinez and Briz, 2000).

Innovation strategy

Studies on the F&B companies' strategies performed in the early 1990s focused on the geography of the market, often in combination with structural and resource-related variables. McGee and Segal-Horn (1992) identified different strategic groups based on two strategy determinants: extent of geographical coverage of the market and brand versus private label orientation. Hyvönen (1993) defined strategic groups on the basis of geographical market differentiation, entry barriers, bargaining power and marketing, production and finance resources. More recent studies address innovation as a central aspect of F&B companies' strategies. Gilpin and Traill (1999) suggest there are different factors determining competitive strategies that are related to company size. Traill (2000) identified eight strategic groups using a wide range of criteria, such as type of competencies, product or process innovation, regional or national focus of brand strategies. Traill and Meulenberg (2002) concluded that patterns of innovation in the F&B industry are more complex than those represented by the traditional theories of 'demand-pull' versus 'technology-push' versus some combination of the two. They suggested that very successful firms have a dominant 'orientation' that permeates the company, forming the company culture and guiding its behaviour. They identified three different orientations: product orientation (focus on product quality), process orientation (focus on issues such as flexibility and efficiency) and market orientation (focus on what the market wants). The firm's dominant orientation demands a set of core competencies (product, process, or market), although successful company will also have to meet basic standards with respect to the other two competences (*ibid.*). Avermeate *et al.* (2004) investigated what are the drivers of product and process innovation in the F&B industry and focussed on SME's. They distinguished between four groups of firms, defined by their innovation strategies: non-innovators, traditionals, followers and leaders. The last three categories include the innovative firms and differ in terms of R&D activities (traditionals have no R&D activities, followers limited and leaders have the most intensive R&D activities). Neither Avermaete *et al.* (2004) or Traill and Meulenberg (2002) used the classification of strategic types introduced by Miles and Snow, although there are common elements.

Innovation strategies give mid-term and long-term direction to product/market/technology combinations (Wheelwright and Clark, 1992). For that reason Miles and Snow's typology is more appropriate for our analysis of the impact of innovation strategies on the product's market performance, than that of Traill and Meulenberg (2002) which focuses on product, or process, or market or that of Avermaete *et al.* (2004) who distinguished types on the basis of R&D intensity. Batterink *et al.* (2006) also used an approach towards innovation objectives which included a combination of product and market-related objectives. They analysed the factors that are related to the innovative output of the Dutch agrifood industry, including explanatory variables related to innovation objectives.

Performance

Several studies have tried to capture the determinants of successful product innovation in the F&B industry (for instance Cabral and Traill, 2001; Earle *et al.*, 2001; Avermeate *et al.*, 2003; Stewart-Knox and Mitchell, 2003; Menrad, 2004; Fuller, 2005). Those focussing on the role of product innovativeness showed that innovative F&B products are more successful on the market than 'copy-cat' or 'me-too' products. Line extensions and most certainly me-too products mostly deliver only short-term, and relatively low-margin benefits (Hoban, 1998; Van Trijp and Steenkamp, 1998; ECR Europe, 1999; Knox *et al.*, 2001). The recent trend of functional foods that addresses specific health conditions even points to the drugs-type features of F&B product development and its related approval and registration regulations (Göransson and Kuiper, 1997; Mark-Herbert, 2004; Stein and Rodríguez-Cerezo, 2008). Companies invest in the development of new products for which they can ask relatively higher prices and which are meant to stay on the market for a long period of time. Joppen (2004) states that if new innovative products catch on, they will almost certainly guarantee long-term commercial benefits in terms of sales and overall profitability. Improved products, such as line extensions and most certainly me-too products, usually deliver only short-term, and relatively low-margin benefits.

Although it can be expected that the most common applied product innovation strategy type in the F&B industry is that of the defender that produces relatively more improved products, it can be assumed that – due to the trends mentioned above – F&B companies increasingly will apply more pro-active strategies by developing relatively more new products. Drawing on the theoretical framework of strategic archetypes of Miles and Snow (1978) and the assumptions presented above that new products will be more successful in the long term and improved products in the short term, we have summarised this in the following hypothesis to be tested in this study.

H3: *The products of F&B companies that intend to follow a prospector innovation strategy will be more successful in the long term than the products from companies that intend to follow another (analyser, defender) innovation strategy.*

4.3 Data and methods

Data on new F&B product introductions have been collected by a systematic review of the issues of 11 different Dutch food trade and professional journals that were published in the second half of 1998. The innovations are identified by sampling the editorially controlled 'new product announcements' sections of these journals; 200 newly launched products were found. For each product we gathered information about the product's name and the name of the company that had developed the product. Additional data including the data that are used for the study presented in this chapter have been collected by a survey using a structured questionnaire.

4.3.1 Variables used in the study

Two dependent variables were used: one for short-term market performance and one for long-term market performance. The set of independent variables included variables for product innovation strategies and strategy-related activities. Three control variables were used: company size, innovativeness of the product and market of the product. See the Appendix for the questions whereupon the variables in this study are based.

Performance of the product

Within the theory of industrial organisations there exist well-established formal measures of company performance (Porter and Scully, 1987; Ferrier and Porter, 1991). Studies on strategy and success have described performance in terms of profitability (Venkatraman and Ramanujam, 1986; Chakravarty, 1986). However, most of the performance indicators are difficult to measure due to unavailability of required data and are often not adequate to predict performance (Bhargava *et al.*, 1994). Hultink and Robben (1995) have investigated how the company's time perspective influences the importance the company attaches to 16 core measures of a new product's success. They found that customer satisfaction was the most important measure regardless of a company's time perspective. For that reason we used customer satisfaction, i.e. market success of the product, as a performance indicator. Short-term market performance could be measured rather specifically; it stands for the financial and market impact of the new product one and a half years after it was announced in the trade journals. The assessment of the financial and market impact was made by the company that introduced the product on the market. Long-term market performance was more difficult to measure; the proxy used was that of the market status of the product seven years after it was announced in the trade journals; either it is on the market, or it is not. Both short-term market performance and long-term market performance are used as dependent variable.

Product innovation strategies and strategy-related activities

Following Miles and Snow (1978), innovation strategies can vary from the prospector type to the defender type, and all combinations between them with the analyser type as a mixture of both. For investigating the company's innovation strategy we used variables measuring the importance of a combination of product-oriented and market-oriented strategic goals for getting engaged in product innovation activities. Three product-oriented strategic goal variables covering the prospector, the analyser and the defender's position have been introduced. Similarly, variables for market-oriented strategic goals have been developed: two for prospector, one for analyser, one for defender. In our definition, companies that follow a specific strategy have a dominant focus on a set of specific strategic goals (product-oriented and market-oriented). Table 4.1 presents the operationalisation of each strategy type, in terms of giving importance to specific strategic goals. However, these are theoretical definitions as companies will also consider other strategic goals (but give less importance to them). The indicators for 'prospector strategy', 'analyser strategy' and 'defender strategy' combine the scores (average) of the importance given to each of the three (for prospector) and two (for analyser and defender) strategic goals.

Market and technology are two main components in a company's innovation strategy; for that reason we included indicators for strategic market assessment and for strategic technology assessment. The strategic market-assessment indicator combines three strategy-related activities the company has (or has not) performed in the process of preparing the product development process: market forecast, competitive analysis and test of product concept by potential clients. The technology-assessment indicator combines three strategy-related activities: the use of internal and of external technologies and the check on patents and licenses. Market-assessment activities are expected to differ between the three strategy types, while technology-assessment activities can be rather evenly applied as technology can be used to implement all strategy types.

Table 4.1. Operationalisation of the three strategy types.

	Give higher importance to the strategic goals:
Prospectors	- extend product range
	- create new markets
	- address specific market needs
Analysers	- replace products being phased out
	- keep up with competitors
Defenders	- improve product quality
	- keep and increase market share

To test the moderating effects, we included two corresponding interaction terms between the relevant independent variable 'prospector strategy' and the two variables for strategy-related activities. To calculate the different interaction terms, we normalised the independent variables prior to computing their cross terms in order to enhance their interpretability and to eliminate non-essential multi-collinearity. To test the interaction between prospector strategy and technology- and market-assessment, we computed the cross-terms 'prospector strategy × market-assessment' and 'prospector strategy × technology assessment'.[5]

Control variables

In line with existing empirical studies on determinants of innovations we included three control variables: company size, product innovativeness and market of the product.

For company size we used a dummy variable: the variable was 1 if the company belonged to the group of large companies (more than 250 employees) and 0 if not. Large firms have relatively more resources and thus can be more innovative in terms of new/improved products and processes than small and medium-sized companies (SMEs). However, there is no consensus in literature on this so-called size effect. Empirical research, seeking for statistical correlation between some measure of innovativity and some measure of either market concentration or company size, produced mixed results over many years (see Cohen, 1995 for a review). Galende and de la Fuente (2003) observed a great empirical disparity; they mention studies that favour a large size (e.g. Horowitz, 1962; Henderson and Cockburn, 1996), but also studies that have not been able to show the benefits of a large size (e.g. Scherer, 1984; Acs and Audretsch, 1988a). Acs and Audretsch (1988b) showed that the contribution of small companies to innovation differs considerably between industries. For the F&B industry Traill and Meulenberg (2002)

---

[5] In a similar vein we calculated cross terms to test the interaction effects of the other strategy variables (analyser and defender) with the technology- and market-assessment variables. As the subsequent regression models including these interaction terms turned out to be insignificant (with these interaction terms not contributing to the model in terms of explanatory power), we do not discuss these interaction terms in the remainder of this chapter.

concluded that only the very small companies innovate significantly less, but for other companies there is no relationship between company size and innovation. However, an empirical study on the agrifood industry did not confirm this (Batterink *et al.*, 2006).

In literature various classifications of innovativeness of products have been proposed. The OECD definition (OECD, 2005) distinguishes between major product innovation (also referred to as radical product innovation) and incremental product innovation. Other classifications use a multi-steps approach indicating several stages of innovativeness (see, for instance, Hermann, 1997; or ECR Europe, 1999). In the present study we used two measurements of innovativeness of the newly announced product: a subjective measure and an objective measure. The subjective measure includes the company's assessment of the innovativeness of the product: is it a new product or is it an improved or renewed version of an already existing product? This variable was used as a dummy variable: new products (1) and improved products (0). The objective measure relates to the number of new product attributes of the product: the more new attributes the higher its level of innovativeness and was used to control the usefulness of the dummy variable. We expected that prospectors would produce more new products and defenders more improved products.

The third control variable deals with the type of market of the product. It is expected that the product-related and market-related strategic goals differ considerably for the type of market the product has to be developed for. For that reason we introduced a dummy variable: we distinguished between products for the consumer market (1) and products for other markets (0). These other markets include the industrial market and the food service market.

4.3.2 Survey

After having tested the draft questionnaire through ten pilot interviews in January 2000 and the redrafting of the questionnaire, data collection took place in the period March-September of 2000. The data on the product innovation strategic goals, the product-related strategy-related activities, the product's innovativeness, the company size and market of the product and short-term market performance have been gathered from phone interviews with managers of the F&B companies that had been directly involved in the product development process. Data on long-term market performance were collected in December 2005, seven years after the product's announcement, mainly through consultation of companies by telephone. Complete data sets have been collected for 129 of the 200 products. Of the 71 products that are not included complete data sets could not be collected for several reasons: the product was not developed in the Netherlands, not brought to the market after all or withdrawn from the market, the responsible person could not be identified or contacted or the company refused to cooperate. Compared to sending questionnaires by post or email – which has an average response rate of 30% – our method led to a relatively high response rate of 76% (129 of 170 eligible targets).

4.3.3 Method of data analysis

The relationships in the conceptual model were tested by means of a regression analysis for measuring the relative importance of the strategy variables for product performance against each other. Linear regression analysis was performed for short-term performance as dependent variable. For long-term performance as dependent variable, a binary logistic regression analysis was used. The predictive power of the regression models was measured by the coefficients of determination (R^2 for linear regression, Pseudo R^2 for logistic regression) which is the proportion of the total variance explained by the model. The t-test (for parametric data) and Mann-Whitney test (for non-parametric data) have been used to test between two groups. T-tests and Mann-Whitney tests provided exact similar results.

4.4 Empirical results

4.4.1 Descriptive results

Nearly two thirds (64%) of the products that were launched in the second half of 1998 are still on the market seven years later, at the end of 2005 (see bottom row in Table 4.2). One quarter of the products with a low performance in 2000, which is one and a half years after product announcement, is still on the market seven years after announcement. Products with a medium short-term market performance have considerable market sustainability in the long term as 67% of them (this is 33% of the total in 2000) are still on the market after seven years; while products with a high short-term market performance have the highest chance of still being on the market after seven years: 74% (this is 28% of the total in 2000).

Correlation analysis between short- and long-term market performance shows a highly significant positive relationship between the two performance indicators (coefficient =0.25; $P<0.01$).

Table 4.2. Short- and long-term market performance for three short-term market performance groups (N=129) (frequencies).

	Short-term market performance	Long-term market performance	
		on the market	off the market
High	49 (38%)	36 (28%)	13 (10%)
Medium	63 (49%)	42 (33%)	21 (16%)
Low	17 (13%)	4 (3%)	13 (10%)
Total	129 (100%)	82 (64%)	47 (36%)

Strategy types and strategy related activities

Companies can use a combination of strategic goals for developing and bringing a new product to the market. Table 4.3 presents the assessment of the importance of the three strategy types and of the specific product-oriented and market-oriented strategic goals that are related to these strategy types. Overall, it can be observed that there are considerable differences between the importance given to the strategic goals: those belonging to the 'Prospector strategy' are considered as most important, followed by that of the 'Defender strategy' with a very high score on the market-oriented goal of increasing/keeping the market share. In general the market-oriented goals are considered as more important than the product-oriented goals. The goal of keeping up with competitors for the analyser strategy is considered of somewhat lower importance.

Table 4.3. Importance of strategic goals and use of strategy-related activities (N=129).

	Means (SD)
Strategy types [1]	
Prospector strategy:	4.03 (0.78)
Extend product range	3.84 (1.43)
Create new markets	3.97 (1.32)
Address a specific market need	4.27 (0.88)
Analyser strategy:	2.66 (1.03)
Replace products being phased out	1.96 (1.39)
Keep up with competitors	3.36 (1.49)
Defender strategy:	3.26 (0.97)
Improve product quality	2.04 (1.61)
Increase/maintain market share	4.47 (0.91)
Strategy-related activities	
Market assessment [2]:	0.84 (0.28)
Marketing research	0.87 (0.34)
Test of product concept by potential clients	0.87 (0.34)
Competitive analysis	0.78 (0.41)
Technology-assessment:	1.55 (0.72)
Use internal technical finding [1]	2.55 (1.43)
Use external technical finding [1]	1.56 (1.08)
Check of patents and licenses [2]	0.55 (0.50)

SD = Standard Deviation.
[1] 5-point Likert scale measuring importance.
[2] Ordinal scale measuring use of activity (0, 1).

Table 4.3 also shows the descriptives of indicators for market-assessment and technology assessment and of the variables on the basis of which they have been constructed. Market forecast activities and testing of the product concept by potential clients were used most often, followed by analysis of competitors. Use of internal technical findings was perceived to be rather moderate importance, that of external technical findings even rather low. Checking of patents and licenses was done in about half of the cases.

Company size, product innovativeness and market of the product

T-tests and Mann Whitney analyses showed some interesting differences between the groups defined by the control variables: company size, product innovativeness and market of the product.

The analyses of the differences between large companies and SMEs show that the large companies in our sample are significantly more focused on following prospector strategies and on developing new products. Also they perform significantly better on market assessment. SMEs in our sample are significantly more focused on using defender strategies (see Table 4.4).

New products (61% of the products in our sample) – as expected – are products with relatively more new product attributes than improved products (39%), although some of the improved products also have a relatively high number of new products attributes. Mann-Whitney tests show that the group of new products differs very significantly ($P=0.00$) from the group of improved products for the variable 'new product attributes' (See Table 4.5).

Table 4.4. Comparison of large companies and SMEs.

	Large companies (N=72) Mean (SD)	SMEs (N=57) Mean (SD)	DbG[1]
New product	0.68 (0.47)	0.53 (0.50)	*
Product for consumer market	0.64 (0.48)	0.53 (0.50)	
Prospector strategy	4.18 (0.68)	3.84 (0.85)	**
Analyser strategy	2.64 (1.09)	2.68 (0.96)	
Defender strategy	3.13 (1.00)	3.42 (0.91)	*
Technical assessment	1.60 (0.65)	1.49 (0.80)	
Market assessment	0.92 (0.19)	0.73 (0.33)	***

SD = Standard Deviation.
[1] DbG - Difference between Groups. Significant difference between the two product groups (using t-test and Mann-Whitney test; excluding cases test-by-test; exact sig 1-tailed) is indicated by *** for $P<0.01$, ** for $P<0.05$ and * for $P<0.10$.

Table 4.5. Comparison of new and improved products.

	New products (N=79) Mean (SD)	Improved products (N=50) Mean (SD)	DbG[1]
New product attributes	4.06 (0.99)	3.54 (1.09)	***
Large company	0.62 (0.49)	0.46 (0.50)	*
Product for consumer market	0.66 (0.48)	0.48 (0.51)	**
Prospector strategy	4.23 (0.67)	3.72 (0.83)	***
Analyser strategy	2.68 (0.96)	2.62 (1.14)	
Defender strategy	3.20 (0.91)	3.35 (1.07)	
Technical assessment	1.59 (0.75)	1.50 (0.69)	
Market assessment	0.84 (0.29)	1.84 (0.26)	

SD = Standard Deviation.

[1] DbG - Difference between Groups. Significant difference between the two product groups (using t-test and Mann-Whitney test; excluding cases test-by-test; exact sig 1-tailed) is indicated by *** for $P<0.01$, ** for $P<0.05$ and * for $P<0.10$.

New and improved products also differ in other aspects. In accordance with the findings above we found that new products are significantly more often produced by large companies and the group of companies that produced these products also applied prospector strategies significantly more often. New products are significantly more often products for the consumer market.

Products for the consumer market differ significantly from the group of products for other markets (industrial market and food service market) in three aspects (see Table 4.6). The group of consumer products consists of significantly more new products as was also found above. The companies that introduced the group of products for other markets are more focused on applying analyser strategies, in combination with defender strategies than those that introduced consumer products.

4.4.2 Product strategy and product market performance

In order to analyse the impact of different product innovation strategies and strategy-related activities on the product's market performance, we introduced two additional indicators that measured the interaction of the prospector strategy (the focus of our analysis) with the two strategy-related activities. A correlation analysis to test for multi-collinearity among the control independent variables is reported in Table 4.7 and based on Spearman's rho. As the largest correlation is (-)0.35 and most of the remaining correlations are significantly lower, multicollinearity is not a significant problem in the analysis (Hair *et al.*, 1998). Furthermore,

Table 4.6. Comparison of products for the consumer and other markets.

	Consumer market (N=76) Mean (SD)	B2B and food service markets (N=53) Mean (SD)	DbG[1]
Large company	0.61 (0.49)	0.49 (0.51)	
New product	0.68 (0.47)	0.51 (0.50)	**
Prospector strategy	4.10 (0.72)	3.92 (0.85)	
Analyser strategy	2.43 (0.95)	2.98 (1.06)	***
Defender strategy	3.12 (0.97)	3.45 (0.95)	*
Technical assessment	1.54 (0.74)	1.57 (0.70)	
Market assessment	0.87 (0.24)	0.79 (0.32)	

SD=Standard Deviation.

[1] DbG - Difference between Groups. Significant difference between the two product groups (using t-test and Mann-Whitney test; excluding cases test-by-test; exact sig 1-tailed) is indicated by *** for $P<0.01$, ** for $P<0.05$ and * for $P<0.10$.

for the linear regression analysis (Model 1) all VIF values are checked (not included in the table); they are all between 1 and 2. This is far below the upper threshold value of 10 (*ibid.*), which also indicates that multi-collinearity problems are not encountered.

Table 4.8 presents the results of the linear regression analysis for short-term market performance (Model 1). Standardised Beta coefficients are displayed (ranging between 0 and 1) for the control variables and for the independent variables (strategy type, strategy activities and interaction of prospector strategy with the two strategy activities). Model 1 is significant ($P=0.088$) and explains 13% of the variance in the short-term market performance. Two variables seem significant in the model explaining short-term market performance. Interestingly, none of the strategy variables are significant.

From this one may conclude that pursuing a specific innovation strategy does not affect the chances for successful (short-term) market performance. However, the model does show that the interaction between prospector strategy and performing market assessment is significantly related to short-term market performance. Apparently, companies that apply a prospector strategy in combination with carrying out extensive market assessments are performing well with their products in the short term. Carrying out technology assessment is a factor for success, and – as technologies can be used to implement all three types of strategies –, is not related specifically to a strategy type. So, as expected a combination of technology assessment with prospector strategy is not a significant factor for successful short-term market performance.

Table 4.7. Correlations diagram for control and independent variables (N=129).

	1	2	3	4	5	6	7	8	9	10
Control variables										
1. Large company	—									
2. New product	0.16 *	—								
3. Product for consumer market	0.11	0.18 **	—							
Independent variables										
4. Prospector strategy	0.20 **	0.33 ***	0.10	—						
5. Analyser strategy	-0.02	0.07	-0.25 ***	-0.12	—					
6. Defender strategy	-0.12	-0.04	-0.14	0.05	0.09	—				
7. Technology assessment	0.13	0.05	-0.03	0.12	-0.03	0.15 *	—			
8. Market assessment	0.35 ***	0.02	0.12	0.25 ***	-0.04	0.24 **	0.21 **	—		
9. Prospector × technology assessment	-0.06	-0.08	0.06	-0.11	-0.15 *	-0.08	0.16 *	-0.08	—	
10. Prospector × market assessment	0.12	-0.05	0.15 *	0.21 **	-0.16 *	-0.09	0.00	-0.02	0.16 *	—

Spearman rank correlations (two-tailed); significant correlations are indicated by *** for $P<0.01$, ** for $P<0.05$ and * for $P<0.10$.

Table 4.8. Regression analyses on strategic goals and activities with short- and long-term market performance (N=129).

	Model I: short-term market performance [1]	Model 2: long-term market performance [2]
Control variables		
Large company	0.00	0.20 (0.47)
New product	-0.02	0.30 (0.47)
Product for consumer market	0.16 *	0.64 (0.47)
Independent variables		
Strategy type		
Prospector strategy	0.12	0.30 (0.34)
Analyser strategy	0.05	0.09 (0.21)
Defender strategy	0.04	0.44 (0.24) *
Strategy-related activity		
Technology assessment	0.17 *	0.41 (0.31)
Market assessment	0.15	0.94 (0.89)
Interaction strategy type and activity		
Prospector × technology assessment	-0.01	0.08 (0.25)
Prospector × market assessment	0.23 **	-0.50 (0.28) *
Constant	2.88 (0.97) ***	-4.05 (2.03) **
R^2	0.13 *	
Adj. R^2	0.05	
Pseudo R^2		0.23 ***
Chi^2 - statistics		23.74
-2 Log likelihood		144.570
F	1.70	
Df	126	

Significant relations are indicated by *** for $P<0.01$, ** for $P<0.05$ and * for $P<0.10$.
[1] Standardised regression coefficients are displayed; for the Constant term the unstandardised regression coefficient and standard error (between parentheses) are displayed.
[2] Beta values (B) and their standard errors (between parentheses) are displayed.

Only one of the control variables – the dummy variable for market type – shows some significant relationship with short-term market performance. So, consumer products perform better in the short term than products that have been produced for the industrial and food service markets.

Model 2 in Table 4.8 shows the results of the binary regression analysis for long-term market performance. Beta values and standard errors are displayed. The model is significant and the value of the Nagelkerke pseudo R^2 which can be interpreted in a similar way to R^2 in linear regression analysis shows some predictive power. The analysis shows that companies following a defender strategy have launched products that are successful in the long term.

Companies that apply a prospector strategy in combination with market assessment – which proved to be a factor for successful short-term market performance – are not successful (significantly negative relationship) in the long term.

In order to analyse the differences between products that were still on the market seven years after introduction and those that has left the market after mid 2000, we made a comparative analysis shown in Table 4.9. Following a defender strategy, but also performing technology assessment and market assessment is significantly more often applied by companies whose products were still on the market seven years after product announcement, as compared to companies whose products were not on the market anymore at that time. Also the products that had left the market are significantly more products for the consumer market, as compared with products for other markets (industrial and food service).

Table 4.9. Comparison of products still on the market and off the market at the end of 2005.

	Still on the market (N=57) Mean (SD)	Off the market (N=72) Mean (SD)	DbG[1]
Large company	0.55 (0.50)	0.57 (0.50)	
New product	0.60 (0.49)	0.64 (0.49)	
Market for consumer market	0.52 (0.50)	0.70 (0.46)	**
Prospector strategy	4.11 (0.70)	3.89 (0.88)	
Analyser strategy	2.73 (1.12)	2.54 (0.84)	
Defender strategy	3.45 (0.99)	2.93 (0.85)	***
Technology assessment	1.65 (0.74)	1.38 (0.67)	**
Market assessment	0.88 (0.26)	0.77 (0.30)	**

SD=Standard Deviation.
[1] DbG - Difference between Groups. Significant difference between the two product groups (using t-test and Mann-Whitney test; excluding cases test-by-test; exact sig 1-tailed) is indicated by *** for $P<0.01$ and ** for $P<0.05$.

4.5. Discussion and conclusions

The contribution of this study lies in the outside-in focus on companies' strategies in the F&B industry. The research question is: what is the impact of the product innovation strategy on the product's short- and long-term market performance? (RQ3).

We investigated the importance companies that had developed the products had given to the three strategy types – prospector, analyser, defender – and the use of strategic technology- and market-assessment activities. Our analyses of the association between strategy types and activities and short and long-term market performance showed that companies that use a prospector strategy in combination with extensive market assessment activities (market forecasting, competitive analysis, test of product concept by potential clients) are especially successful in the short term, but unsuccessful in the long term. Companies following a defender strategy are most successful in the long term. Surprisingly, the analyser strategy did not prove to be a successful innovation strategy for F&B companies. For that reason we must conclude that Hypothesis 3 cannot be confirmed. This hypothesis stated that the products of F&B companies that intend to follow a prospector innovation strategy will be more successful in the long term than the products from companies that intend to follow another (analyser, defender) innovation strategy.

One conclusion could be that companies focussing on a prospector strategy – aiming at expending the product range and creating new markets, also by addressing specific market needs – while thoroughly preparing the new product development process (through market assessments) are the most successful in gaining short-term market performance of F&B products. Companies following this strategy maintain a high level of innovative activity by regularly introducing new products to the market in the period to 2006, thereby replacing the 'old new products' we found in 1998. On the contrary, once companies following a defender strategy have introduced a new product, they are less likely to engage again in developing new products, replacing the older ones. Instead, they will try to keep their products on the market for as long as possible. Companies following a defender strategy are able to maintain – by regular small improvements – and protect the position of their once introduced product on the market on a certain level of successful market performance. In other words, a first mover's strategy must be refocused for sustaining the market for the product for the long term. However, if all companies followed this best strategy they would become less successful in the end while companies that prefer to choose an alternative strategy might profit from attractive market segments. In practice, companies have a product portfolio and their products under development are in several stages of development; so most companies follow a combination of strategies depending on the innovativeness of their products and the phase of the life cycle their products are in.

Although Mann *et al.* (1999) found that companies in the F&B industry – on average and compared to a number of other industries (aerospace, insurance, automotive) – are weak in policy and strategy, our results show that companies that have formulated product innovation strategies and give them high importance are successful. It is important for firms to have a clear strategy as this will direct the process of competence formation, of developing linkages with the outside world and of deciding which types of innovation the company concentrates upon. In other words: strategy matters. Overall, our results confirm that formulating an explicit product innovation strategy is a key to a F&B product's successful short- and long-term market performance. For the short term this applies to a prospector strategy in combination with market assessment activities and for the long term this applies to a defender strategy. Traill and Meulenberg (2002) in their study on the factors for successful product innovation in the European F&B industry found that describing a company's innovation strategy in terms of only one dominant orientation (product, process or market) is not sufficient, as in reality a finer segmentation of company types might be more appropriate. We applied a more detailed segmentation as we operationalised innovation strategy as a combination of market- and product-oriented goals (see Table 4.3). Our results showed that a distinction should also be made between short- and long-term innovation strategies. This could add to an even more appropriate set of variables for measuring innovation strategy.

To date our study is the first that has provided empirical evidence on the long-term market performance of products in the F&B industry and on the importance of companies having a clear innovation strategy for achieving successful short and long-term market performance. This is also one of the first quantitative studies focusing on strategic objectives and subsequent NPD activities related to the innovative output of the Dutch F&B industry. Batterink et al. (2006) included this sector in their study on the agrifood industry, but did not distinguish between the two – agriculture and F&B industry – although their sectoral innovation systems can differ considerably (Malerba, 2005). Although some explorative studies also include innovation strategies in the F&B industry (Traill, 2000; Traill and Meulenberg, 2002; Avermate et al., 2004; Batterink et al., 2006) this is also the first study that focuses on products and not on the company or the sector level. By linking multiple indicators for innovation strategy to specific product innovations and the product's market performance, this study contributes to a broader understanding of the strategy-related factors that contribute to the product's success on the market.

The research contains some weaknesses as well. The proxy variables used in the hypothesis test are crude representations of the underlying concepts and can be thought of as providing no more than limited support for the hypothesis. This applies for instance for the variable that measured long-term market performance of the products. We have opted for an ordinal variable that measured whether the product that was introduced in 1998, was still available on the market seven years later. We would recommend developing a more segmented variable for measuring long-term market performance similar to that for short-term market performance. Moreover, it can be expected that companies have been monitoring their products over the years and have implemented small adjustments to the products. We therefore suggest that further research be conducted that investigates in more depth whether and on what scale products that have been introduced as new to the market evolve over the years. Such a study should also investigate how regularly monitoring of market performance influences this and what impact the various factors involved in these adjustment processes (strategy-related, but also others) can have on the long-term market performance of the new product. If the product is part of a product group of a company, further research could be conducted on the effect this newly launched product has on the existing products as companies will create their own benchmark.

5. Innovation network: open innovation

5.1 Introduction

This chapter presents the results of the study on the openness and composition of the external networks involved in the innovation process of new versus improved F&B products.[6] It investigates the role of specific external resources on the product's short and long-term market performance.

Innovative companies try to reach and maintain competitive advantage by developing products that bring them a sustainable position on a market that is characterised by international competition and increasing customer demands. Speed has become an important competitive weapon (Cooper, 1993). In order to speed up the new product's innovation process, innovative companies increasingly use ideas and resources from outside the company. Volberda *et al.* (2006, p. 10) have phrased this by stating that 'innovative companies know their weaknesses. That is one of their strengths'. As a consequence of this, networks have become vital for an innovating company's strategy for survival and growth. Therefore one of the most critical questions to be answered by the management of innovating companies is: in what role and for which activities can we successfully involve other companies and organisations in our innovation process?

Up to now most studies on innovation networks and the success of innovation deal with the company as a whole and not with a specific product that is under development. In addition, hardly any research has been done on the development of products that stay on the shelf for a longer period of time, i.e. products that have become the 'cash cows' of the company. The present study aims to fill these gaps: we investigate the relationship between the composition of the external network of the company that has developed the product and the product's market performance, not only soon after market launch, but also after several years.

The main research question of the study is:

RQ4: *What is the impact of the openness and composition of the product-related innovation network on the short- and long-term market performance of the product?*

Chesbrough (2003) introduced the concept of 'open innovation': it refers to the phenomenon that innovative companies increasingly use resources from outside the company to speed up the innovation process. Not only R&D-intensive firms, but also less R&D-intensive firms rely on external resources as their infrastructure is not sufficient to innovate on their own.

---

[6] This chapter is based on: The impact of the openness of the innovation process on the short-term and the long-term market performance of new products. Evidence from new product announcements of the Dutch food and drinks industry, Enzing, C.M., Janszen, F.H.A. and S.W.F. Omta (2008). Paper presented at the 8th International Conference on Management in AgriFood Chains and Networks, 28-30 May 2008, Ede, the Netherlands.

Nevertheless, most empirical studies on open innovation deal with high-tech industries such as biopharmaceuticals, ICT and computers (e.g. Christensen *et al.*, 2005; Fetterhoff and Voelkel, 2006; Dittrich and Duysters, 2007) and with a strong focus on large and predominantly US-based firms (Chesbrough, 2003, 2006).

Although there is a widespread practice of cooperation between companies within the food value chain, empirical investigations on open innovation in low- and medium-tech industries, such as the F&B industry are scarce (Huston and Sakkab, 2006; Vanhaverbeke and Cloodt, 2006; Smit *et al.*, 2008; Sarkar and Costa, 2008). That open innovation might be very interesting to companies in the F&B industry is indicated by the finding of Archibugi *et al.* (1991) that F&B firms rely even more on external resources than the average for all industries. This study focuses on open innovation in the Dutch F&B industry. Within the Netherlands, it is the largest manufacturing sector (representing 22.4% of the total turnover of all manufacturing industries) and the leading employer representing 20% of the total Dutch manufacturing industry (LEI, 2008; data for 2005). It is also an important player in Europe: the Dutch contribution to the European (EU25) total turnover in this industry was 6.3%, while the Dutch F&B companies only constitute 1.8% of the total number of F&B companies in EU25 (INNOVA, 2008).

We have built a database of 129 F&B products that holds data on products which have been developed with and without using an external network. The market launch of the products was announced in the second half of 1998. Data have been collected on the product's innovativeness, on the involvement of (supply chain) companies and public research organisations in the idea generation stage and in the product development stage of the innovation process. Their performance was first measured in the beginning of 2000 and again at the end of 2005, respectively one and a half years and seven years after their announcement.

The chapter is structured as follows. Section 5.2 presents the theoretical framework of the study and the hypothesis to be tested. In Section 5.3, we briefly describe the methods of data collection, the operationalisation of the variables and the data analysis. Section 5.4 presents the empirical results of the study. The last section discusses the main findings and draws conclusions.

5.2 Theoretical framework and hypothesis

5.2.1 Innovation

Innovation, as Schumpeter (1939) defined it, is 'any doing things differently in the realm of economic life'. So essentially, innovation is about change: change in the products a company makes and change in the way the company produces them, also referred to as product innovation and process innovation (Tidd *et al.*, 2005). There are degrees of change; from only minor incremental improvements, adaptations or refinements of existing products and processes to very radical changes leading to totally new products or production processes. Product and process innovations are the result of an interactive process in which actors within the innovating company together with actors from other organisations transform knowledge and techniques into new products and processes. Knowledge can be existing or new scientific and technological knowledge, knowledge of (new) markets and of organisations (McKelvey, 1996). Rothwell (1992) has described how our understanding of the innovation process has evolved from a simple linear model in which innovation was basically science driven to increasingly complex interactive models in which the need for cross-functionality across the company's borders was recognised. The most updated model considers the innovation process as an interactive, cumulative and co-operative phenomenon in which actors from inside and outside the company participate and which requires high levels of integration at both intra- and inter-firm levels (Teece *et al.*, 1997). Success and competitive advantage depend on the ability of the company to integrate, build and reconfigure internal and external resources to address rapidly changing environments (*ibid.*).

5.2.2 Open innovation

In 2003, Chesbrough introduced the concept of 'open innovation' to describe these changes and more specifically how companies have come to use external ideas and resources in order to speed up the innovation process. He argues that in contrast to the old model of 'closed innovation' in which large R&D-intensive companies conduct their R&D solely in-house, these companies have become more open in their innovation processes. The concept of 'open innovation' describes the phenomenon that large knowledge-intensive companies increasingly acquire knowledge (R&D) externally. Even the largest companies cannot rely only on internal resourcing; they also require knowledge from outside when developing innovations (Rigby and Zook, 2002; Chesbrough and Crowther, 2006). The ability to combine internal and external information is a critical new source of competitive advantage (Rigby and Zook, 2002). Cassiman and Veugelers (2006) suggest that external inputs can increase the productivity of in-house activities. Key drivers for opening up the innovation process include the increasing availability and mobility of knowledge workers, the flourishing of the venture capital market and the increasing scope of capable external suppliers (Chesbrough, 2003).

But not only large and R&D-intensive companies cooperate with external knowledge actors. Smaller companies and companies in low- and medium-tech industries also operate in knowledge networks. They are even more dependent on the contribution of and cooperation with external actors as their innovation-related infrastructure is not sufficient to innovate on their own (Hirsch-Kreinsen *et al.*, 2005). Evidence reveals that the open innovation concept and associated strategies is used also in more traditional and mature industries (Chesbrough and Crowther, 2006; Huston and Sakkab, 2006; Smit *et al.*, 2008). Collaboration with external actors improves the company's cooperative competences; most innovation studies found that these were positively associated with a product's performance (Sivadas and Dwyer, 2000; de Man and Duysters, 2005).

Overall, open innovation emphasises the need for companies to network with other actors throughout the innovation process. Open innovation combines a number of trends that scientists have recognised already for a long time, including the role of lead users (Pavitt, 1984; Von Hippel, 1988), innovation networks (Lundvall, 1992; Hakansson, 1995) and the interactive, cross-disciplinary and (mostly) inter-organisational nature of the innovation process (Kline and Rosenberg, 1986).

5.2.3 Open innovation in the F&B industry

In the F&B industry the interactions between companies and their business partners in the supply chain as well as with (public) research organisations play a crucial role in achieving successful innovations. Archibugi *et al.* (1991) found that F&B companies rely more on external sources of innovation than the average for all industries. Companies in the F&B industry have developed a broad interface with innovators in other industries and apply scientific advances that have been developed in these other industries (Christensen *et al.*, 1996; Rama, 1996; Knudsen, 2007). As the F&B industry has to operate on a buyer's market, market-orientation is considered as a key success factor for innovation in this industry (Grunert *et al.*, 1996; 1997, 2008; Kristensen *et al.*, 1998; Omta *et al.*, 2003; Batterink *et al.*, 2006; Sarkar and Costa, 2008). That is why F&B companies have developed networks with actors that provide them with market intelligence; through these networks they keep track of their end-users and explore future consumer trends.

Knox *et al.* (2001) found that wide consultation with agencies and the involvement of expertise beyond the company had a positive impact upon the success of F&B products.

In this book we use a sectoral systems approach towards innovation processes; this approach focuses on all relevant actors and institutions that are involved in the innovation process of an industrial sector (Breschi and Malerba, 1997). More specifically for the study presented in this chapter we focus on the external actors that are involved in the F&B companies' product innovation processes: the product-related innovation network. We make a distinction between those actors in the product-related innovation network that provide ideas for new or improved products and those that are involved in the development of these products.

We expect F&B companies that innovate in networks to be more successful than companies that do not. However, it should be realised that too large a network of alliances may lead to saturation and overembeddedness (Kogut *et al.*, 1992; Uzzi, 1997). When the company is involved in a too dense network, it can limit a company's openness to information and flexibility to operate (Nahapiet and Ghosdal, 1998). The management of the different network links and the overall coordination of all these linkages require a lot of attention and costs can increase significantly (Harrigan, 1985). Although there will be a limit to the number of external relations that can be managed by a company successfully (Gomes-Casseres, 1996), we expect that this restriction will not apply to F&B companies. Although the F&B industry might use a relatively high number of external resources (Archibugi *et al.*, 1991), it is expected that most of them do not demand active relationship management as they might deal with goods that are bought. When it comes to R&D partnerships which demand active relation management, Hagedoorn (2002) found that in the period 1960-1998 the share of newly established R&D-partnerships of low-tech industries such as food and beverages, metals, oil and gas was relatively small compared to high-tech sectors such as aerospace, pharmaceuticals and the information technology industry. Summarising the considerations above, we have formulated the hypothesis of the study as follows:

Hypothesis 4: *The more open the product-related innovation network, the better the product's short-term and long-term market performance.*

We expect to find differences in the level of involvement of specific external actors in the product's network and their impact on the short-term and long-term market performance, depending on the innovativeness of the product. We have elaborated this in more detail in the next section.

5.2.4 Involvement of technology-related and market-related actors

The use of new technologies has become a main factor that explains the differences between a line extension and a truly new F&B product (Katz, 1998). Traill and Meulenberg (2002) found – on the basis of a survey of European food-manufacturing companies – strong evidence that R&D expenditure was closely correlated with the development of new products. This illustrates the increasingly science- and technology-based character of the innovation process in the F&B industry. However, as F&B companies have only limited resources for scientific and technological activities, it is expected that F&B companies become more dependent on external knowledge inputs when developing more innovative products. This can be input from technology-related actors, such as research organisations (universities, research institutes, polytechnics), consulting agencies and companies that provide, goods and related services (such as training). Omta *et al.* (2003) found that research organisations are important sources of innovation in the agrifood industry, especially in the earliest stages of the innovation process. F&B companies can benefit from new scientific and technological developments that have been developed within these research organisations and by other companies and that are embodied in the products and services of these actors. Through a well-developed network of inter-industry purchases of machinery and equipment, as well as raw materials and ingredients, F&B companies can use these new technological developments (Klevorick *et al.*, 1995; Christensen *et al.*, 1996; Galizzi and Venturini, 1996; Traill and Meulenberg, 2002). Technological consultants were found to be very relevant for companies that develop new products but are too small to invest in in-house R&D, especially in mature industries (Smallbone *et al.*, 1993). In addition, the role of training was found to be important, especially in low- and medium-tech industries because many employees need to have knowledge of several disciplines (Schmierl and Köhler, 2005). Bringing the company employees knowledge up to date, thereby increasing its absorptive capacity (Cohen and Levinthal, 1990) is also related to innovation performance (Warner, 1996).

Those companies that provide goods and (related) services can operate within (raw materials, ingredients) or outside (equipment, machinery, recipe consultancy) the supply chain. As innovations in the F&B sector are becoming more knowledge intensive this might have an impact on the role of these companies in the innovation process. Petroni and Panciroli (2002) found that companies assign supply companies different roles and give them varying levels of responsibility in the product development process. These roles are correlated to the supplying companies' distinctive innovation capabilities. For the F&B industry this might imply that these companies may not only sell their goods and services, but that they play a more active role, for instance as partner with whom F&B companies collaborate in the innovation process or to whom specific tasks in the innovation process are outsourced. See, for instance, Joppen (2004) on the role of the ingredients companies: they proved to be of growing importance in the food innovation processes as they provided important in-depth information on intrinsic aspects of ingredients (such as flavours, fragrances, antioxidants for health benefits, fat and sugar substitutes). When distinguishing between new and improved products, it might be expected that in the case of new products these supplying companies will be involved relatively more often in the role of partner or of outsourcer and in the case of improved products in the role of sellers of technology-embodied goods and services. For both product groups the involvement of technology-related external actors is expected to be crucial for successful market performance in the short term; for new products – due to larger investments that have to be earned back during a longer period of time – also in the long term.

F&B companies that bring new products to the market also do this on the basis of extensive market knowledge, partly based on the input from different market-related actors. In doing so, companies have a better view of the long-term market needs and the positioning of the new brand. Market-related actors can be companies that provide market knowledge such as those performing market research or that advise on brand (re)positioning, but also customers i.e. clients in the business-to-business market, consumers and competitors. Involving customers in the innovation process is associated with retrieving market information, while involving competitors is associated with exploiting economies of scale and reducing individual costs of innovation (Miotti and Sachwald, 2003). Customers are an especially important source of innovation for companies in the F&B industry (Enzing *et al.*, 1996; Omta *et al.*, 2003). In the F&B industry products of competitors are objects that can be easily copied or imitated (as their IP is not protected); this makes them an interesting source of new product ideas. The smaller F&B companies in particular were found to be using this strategy: the most innovative firms in this category where indicated as followers (Avermaete *et al.*, 2004). New products need advanced and carefully prepared marketing efforts in order to be recognised by the public. In the case of improved products it can be argued that the market is already known and the involvement of external market-related actors for providing market knowledge is less crucial.

5.3 Data and methods

5.3.1 Data collection

Data on F&B product introductions were collected by a systematic review of the issues of 11 different Dutch food trade and professional journals that were published in the second half of 1998. We screened the Dutch food trade and professional journals until we had identified 200 new products. For each product we gathered information about the product's name and the name of the company that had developed the product. Additional data, including the data that are used for the present study, have been collected in a survey using a structured questionnaire. After having tested the draft questionnaire through ten pilot interviews in January 2000 and the redrafting of the questionnaire, data collection took place in the period March-September of 2000. The data on the product's innovativeness, the product-related innovation network and short-term market performance have been gathered from phone interviews with managers of the companies that have been directly involved in the product innovation process. Data on long-term market performance were collected in December 2005, seven years after the product's announcement, through telephone interviews with the companies.

Complete data sets were collected through the survey for 129 of the 200 new products, in 2000. Of the 71 products that are not included 45% were not eligible (not developed in the Netherlands, not brought to the market after all, withdrawn from the market) and for the other 55% data could not be collected for several reasons (responsible person could not be identified or contacted, company refused to cooperate). Data collection on long-term market performance in 2005 could be completed for all 129 cases.

5.3.2 Variables

The variables used in this study include: openness of the product-related innovation network, innovativeness of the product and performance of the product. See the Appendix for an overview of the operationalisation of the variables that are used in the present study and the related questions used in the survey.

Openness of the product's innovation network

In the present study we focus on the product-related innovation network at the start (in Coopers' terminology: 'the ideation') and in the development stage of the product innovation process. In the idea stage, external actors are involved as a source of ideas. In the product development stage they can be involved in three different roles: as partner, outsourcer or as seller (see Table 5.1 for definitions of roles of external actors).

We distinguish between technology-related and market-related actors (see Section 5.2.4). For the group of technology-related actors we investigated the role of public-funded research

Table 5.1. Roles of external actors in the product innovation process.

Idea stage	
Source of ideas:	external actors can provide ideas for new products (constituents) or their making
Product development stage	
Partner:	companies or other organisations that are directly involved in the company's product innovation process
Outsourcer:	companies or other organisations to which specific activities in the product innovation process are outsourced
Seller:	companies or other organisations from which the company buys specific goods and services that are related to the product innovation process

organisations (universities, research institutes and polytechnics) as source of ideas, partner and as outsourcer. Companies that develop and produce technology-embodied products and (related) services for the F&B industry are studied in four different roles: as source of ideas, as partner, outsourcer and as seller. For the group of market-related actors, customers and competitors are included both as source of ideas and as partner. Companies providing market-related knowledge are included in their role as outsourcer. The openness of the network is operationalised using the number of different external actors that are involved in the idea stage and in the product development stage. The higher the number of different external actors involved, the more open the network.

Innovativeness

In literature various classifications of innovativeness of products have been proposed. The definition in the Oslo Manual (OECD, 2005) distinguishes between major product innovation (also referred to as radical product innovation) and incremental product innovation. Other classifications use a multi-step approach indicating several stages of innovativeness (see for instance Herrmann, 1997; or ECR Europe, 1999). In the present study we have used two measurements of innovativeness of the product: a subjective measure and an objective measure. The subjective measure includes the company's assessment of the innovativeness of the product: is it a new product or is it an improved or renewed version of an already existing product? The objective measure relates to the number of new product attributes: the more new attributes, the higher the product's level of innovativeness.

Performance

Innovation performance has been operationalised in many ways. Cooper and Kleinschmidt (1987) found three independent dimensions that characterise product performance: financial performance, opportunity performance and market impact. Financial and market performance seem most suitable when the performance of a product has to be measured. A variety of financial accounting-based and market-related indicators of performance can be used (see for a review Murphy *et al.*, 1996).

On the basis of a survey of 300 food processing companies, Kristensen *et al.* (1998) found that – from a set of eight different success criteria – the impact on the company's market share and on the company's earning capabilities are by far the best indicators to measure the success of new food products. Tidd *et al.* (2005) argue that the time perspective needs to be considered: success in the short term 'might be a result of a lucky combination of new ideas and receptive market at the right time' (p. 37) but this is not enough for a company that aims at sustainable growth over a long period of time. For products this includes growth of sales, growth in turnover, etc.

We used two performance indicators. Short-term market performance stands for the financial and market impact of the new product one and a half years after product launch. As there are no objective financial or market data available on individual F&B products, subjective sources had to be used. In our case the assessment of the financial and market impact was made by the company that had introduced the product on the market. Long-term market performance stands for the market status of the product seven years after it was announced in the trade journals.

5.3.3 Method of data analysis

The relationships in the conceptual model were tested by means of correlation analyses. The Spearman rank correlation coefficient for non-parametric data was used to measure the significance of the relationship between short-term market performance and indicators for the level of openness of the innovation network. Missing values were deleted listwise. Chi-square statistics (Phi coefficient for 2x2 contingency tables, Cramer's V coefficient for 2x(2+n)) were used for measuring the strength of the associations between long-term market performance and the indicators for the level of openness of the innovation network. We also used single linear regression for short term and single binary logistic regression for long-term market performance: this did not alter the conclusions drawn on the basis of outcomes of the correlation analyses. The Mann-Whitney test is used to test between two groups and is suitable for non-parametric data analysis. The outcomes of Mann-Whitney tests (2-Independent samples test) with long-term market performance as grouping variable did not alter the conclusions drawn on the basis of outcomes of correlation and regression analysis dealing with long-term performance.

5.4 Empirical results

5.4.1 Baseline description

The 129 products represent a large variety of product groups of the F&B industry: ready-to-made or use meals (including breakfast products) and snacks; processed fruit and vegetable products; dairy products; grain mill, starch, bakery and farinaceous products; beverages; chocolate and sugar confectionery and ingredients for food products. Of these 129 products, 61% are new products and 39% are improved products. Most new products are in the ready-to-make product group (80%), the food ingredients product group (73%) and the grain mill, starch, bakery and farinaceous product group (63%). The smallest number of new products is in the chocolate and sugar confectionery group (40%). In the other three groups (fruit and vegetables, dairy and beverages) the ratio between new and improved products is about fifty-fifty.

Product innovativeness

Table 5.2 shows for the new and improved product groups the distribution across the variable which measures product innovativeness in terms of number of new product attributes. New products – as expected – are products with relatively more new product attributes than improved products, although a few of the improved products also have a relatively high number of new attributes.

The Mann-Whitney test shows that the group of new products differs very significantly ($P=0.00$) from the group of improved products for the variable 'new product attributes'.

Composition of the product-related innovation networks

The external network of F&B companies can provide important sources of ideas for product innovations. Ideas are the feedstock of the new product's development process (Cooper, 1993). Our results show that for the whole sample market-related actors – customers and

Table 5.2. Newness of the product and new product attributes (frequencies, N=129).

Newness of the product		Number of new product attributes							
		0	1	2	3	4	5	>5	Total
New products	61%	0%	25%	56%	13%	2%	4%	1%	100%
Improved products	39%	10%	50%	26%	6%	6%	2%	0%	100%
Total	100%	4%	35%	44%	10%	3%	3%	1%	100%

competitors – are particularly important sources of innovation (see Table 5.3). Companies that supply specific goods were also used frequently as a source of ideas: those selling raw materials and ingredients more often than those selling equipment or machinery. Research organisations (universities, research institutes, polytechnics) have hardly been used as external sources of ideas.

The most frequently used partners in the development stage of the product innovation process were companies supplying ingredients, machinery and equipment and customers. Research organisations and competitors also operated as partners, but less frequently. Outsourcing was carried out most frequently for market-related activities. Research organisations and

Table 5.3. Involvement of external actors in innovation networks (frequencies).

Actors in external network	Whole sample	New products	Improved products	DbG[1]
Idea stage				
Technology-related				
Research organisations	2%	3%	2%	
Ingredients companies	16%	14%	18%	
Equipment companies	5%	6%	2%	
Market-related				
Customers	37%	29%	50%	**
Competitors	24%	25%	22%	
Product development stage				
Technology-related				
Partner - research organisations	8%	8%	8%	
Partner - supplying companies	35%	39%	28%	
Outsourcer - research organisations	10%	9%	12%	
Outsourcer - company providing recipe advice	5%	5%	4%	
Seller - equipment companies	29%	26%	35%	
Seller - training companies	14%	14%	14%	
Market-related				
Partner - customer	35%	39%	28%	
Partner - competitor	4%	4%	4%	
Outsourcer - marketing companies	40%	47%	29%	**

Whole sample, N=129; New products N=79; Improved products N=50.
[1] DbG - Difference between Groups. Significance of difference between the two product groups (using Mann-Whitney analyses; excluding cases test-by-test; exact sig 1-tailed) is indicated by ** for *P*<0.05.

companies that provide recipe advice were less frequently involved for outsourcing activities. In more than one quarter of the cases companies that produce machinery and equipment were involved as sellers in the product development stage. Furthermore, education and training activities were provided by external organisations.

We also investigated the differences in the composition of the product-related innovation networks between new and improved products in the idea and development stages (Table 5.3, last two columns). For the new products we found that in the idea stage relatively – but not significantly – technology-related actors have been involved more often. This does not apply to the suppliers of ingredients and raw materials; they might have provided ideas for the new ingredients that could have replaced existing ingredients in the improved product. Also in the idea stage customers are significantly more often a (market-related) source of innovation for improved products. Competitors are relatively more often a source of innovation or new products.

The composition of the innovation networks in the development stage also shows differences between the two product groups. In the case of new products, the companies work together with partners more often, especially with companies that supply ingredients, machinery and equipment and with customers, but the differences are not significant. Outsourcing of market-related activities took place significantly more often in the case of new products. Outsourcing to research organisations was performed relatively more often for improved products. Also companies selling machinery and equipment were more often involved in the improved product's networks.

The reader should bear in mind that the density and frequencies of interactions has not been measured. For instance, a company might have several different companies that are customers and have operated as a source of innovation for the specific product, but they have been included as one type of external actor in this role.

Short- and long-term market performance

The results show that nearly two thirds (64%) of the products that were announced in the second half of 1998 are still on the market seven years later, at the end of 2005 (see bottom row in Table 5.4). Products were grouped according to their market performance in the short term (measured one and a half years after product announcement) in three categories: low (13%), medium (49%) and high (38%). Three quarters of the products with a low market performance in the short term (one and a half years after product announcement) have disappeared from the market seven years after announcement. Products with a medium short-term market performance have much larger market sustainability in the long term (67% of 49%); while products with a high market performance in the short term have the highest chance of a long-term market position (73% of 38%). Similar conclusions can be drawn for the new products and improved products separately. There are only small and not significant differences between

Table 5.4. Short- and long-term market performance (frequencies).

	Short-term market performance			Long-term market performance		
	Whole sample	New products	Improved products	Whole sample	New products	Improved products
High	38%	43%	31%	73%	74%	73%
Medium	49%	43%	59%	67%	62%	72%
Low	13%	14%	10%	25%	27%	20%
Total	100%	100%	100%	64%	62%	67%

All products N=129; New products N=79; Improved products N=50.

the product groups: in the short term new products show better performance and in the long term improved products perform better.

Correlation analyses confirm the relationship between short and long-term market performance. For the whole sample there is a highly significant positive relationship between the two performance indicators (coefficient=0.32; P= 0.00; df =2) and significant for the group of new products (coefficient=0.31; P=0.02; df =2) and improved products (coefficient=0.34; P=0.06; df =2).

We also investigated the relationship between product innovativeness and market success by means of correlation analyses between the product innovativeness variable dealing with the number of different product attributes and the two performance indicators. We found a significant positive relationship between the number of different product attributes and long-term market performance (coefficient =0.19; P=0.09; df =2), but not for short-term market performance (coefficient =0.02; P=0.41). So product innovativeness is a factor for sustainable market success: the more new product attributes the better the long-term performance of the product. Apparently, the market is able to recognise and appreciate the new qualities of the product.

5.4.2 Level of openness and composition of the network and performance

Finally we investigated the role of the openness of the product-related innovation networks and more specifically the role of each of the technology-related and market-related actors in the products' short- and long-term market performance.

Openness and market performance

Our results show that openness of the product-related innovation network is significantly positively related to the product's market performance: the more different actors involved the better the short- and long-term market performance (see Table 5.5). Openness of the innovation network in the idea stage (in terms of number of different sources of ideas) relates positively to both short- and long-term market performance; the most significant to the latter. Openness of the innovation network in the development stage (in terms of the number of different partners, outsourcers and sellers) relates significantly positive to both performance indicators; most significant to short-term market performance. Correlation analyses for each of the three types of actor groups (partners, outsourcers, sellers) separately with the two performance indicators (not presented in the table) show that only the number of different sellers relates significantly positively with short-term market performance (coefficient =0.21; $P<0.01$).

When investigating the role of networks openness in the market performance of the two product groups separately (Table 5.5) we found that openness of the new product's innovation network in the idea stage is highly significant and positively related to the new product's long-term market performance. In the development stage of new products, network openness is significantly positively related to both performance indicators; most significantly to short-

Table 5.5. Relationship between openness of the product-related innovation network and the product's short- and long-term market performance.

Openness	Short-term market performance [1]	Long-term market performance [2]	
		Phi/Cramer's V	df
Idea stage			
Whole sample	0.14 *	0.33 ***	4
New products	0.13	0.47 ***	4
Improved products	0.18	0.26	4
Product development stage			
Whole sample	0.16 **	0.28 *	5
New products	0.19 **	0.35 *	5
Improved products	0.09	0.23	5

Whole sample N=129; New products N=79; Improved products N=50.
Significant correlations are indicated by *** for $P<0.01$; ** for $P<0.05$ and * for $P<0.10$.
[1] Spearman coefficient, one-tailed.
[2] Chi-square coefficient: Phi for 2×2 contingency tables, Cramer's V for 2×(2+n); df – degrees of freedom.

term market performance. For the group of improved products, openness of the innovation network does not play a significant role in the improved products' short-term or long-term market performance either in the idea stage or in the development stage of the product innovation process.

So for the successful and sustainable development of new F&B products the involvement of many actors, both as sources of innovation and in the products development stage is of crucial importance. For improved products, their involvement is not of crucial important (anymore).

The role of technology-related and market-related actors in the new and improved product's market performance

Finally, we investigated the relationship between the involvement of each of the technology-related and market-related actors separately with short and long-term market performance for the two product groups (Table 5.6). For the group of new products we found that in the idea stage, technology-related actors that contribute significantly to the products' short-term success are the suppliers of machinery and equipment. In the development stage the involvement of these actors in the role as seller of the machinery and equipment relates significantly positively to the new product's short-term success. This also applies to the involvement in the network of companies that have provided training. The outsourcing of market-related activities has a significantly positive impact on short-term market performance of new products.

Customers and competitors as sources of innovation have contributed significantly to the long-term market performance of new products: companies that use both market-related sources of innovation have better long-term market performance than companies that do not use them. The high significance of the involvement of customers as source of innovation has been found in many experimental studies: see for instance Von Hippel (1988) on the important role of lead users who help improve the product and reduce costs, and Oudshoorn and Pinch (2003) on how users co-construct technology. However, the involvement of customers as partner in the product development stage of the innovation process does not show any significant relationship to the new product's short or long-term market performance.

In the development stage of new products we found that the involvement of research organisations to which activities have been outsourced and of companies from which machinery and equipment was purchased is of significant importance.

Additional analyses (results are not included in the table) of each of the three research organisations separately – universities, research institutes, polytechnics – shows that only the involvement of research institutes relates significantly positively to long-term market performance (coefficient of 0.20; $P<0.10$).

Table 5.6. Relationship between involvement of external actors and the product's short- and long-term market performance.

Actors in external network	Improved products			New products		
	Short-term market performance[1]	Long-term market performance[2] Phi/Cramer's V	df	Short-term market performance[1]	Long-term market performance[2] Phi/Cramer's V	df
Idea stage						
TR Sol - research organisations	-0.05	0.10	2	0.07	0.13	2
Sol - equipment companies	-0.05	-0.20	1	0.16 *	0.20	1
Sol - ingredient companies	-0.12	-0.32 **	1	0.06	0.01	1
MR Sol - customers	-0.05	0.13	1	0.07	0.33 ***	1
Sol - competitors	-0.02	-0.03	1	0.11	0.28 **	1
Product development stage						
TR Partner - research organisations	-0.12	0.20	2	0.10	0.22	2
Partner - supplying companies	0.08	-0.31 **	1	0.11	-0.12	1
Outsourcer - research organisations	-0.03	0.01	1	0.07	0.24 **	1
Outsourcer - company providing recipe advice	0.12	-0.07	1	0.01	-0.06	1
Seller - equipment companies	0.09	-0.01	1	0.25 **	0.21 *	1
Seller - training companies	0.02	-0.07	1	0.27 ***	0.16	1
MR Partner - customer	0.09	0.18	2	0.01	0.09	2
Partner - competitor	-0.02	0.15	2	-0.05	0.10	2
Outsourcer - marketing companies	0.05	-0.01	1	0.20 **	-0.05	1

New products N=79; Improved products N=50.

TR = Technology-related; MR = Market-related; Sol = Source of ideas.

Significant correlations are indicated by *** for $P<0.01$; for ** $P<0.05$ and * for $P<0.10$.

[1] Spearman coefficient, one-tailed.

[2] Chi-square coefficient: Phi for 2×2 contingency tables, Cramer's V for 2×(2+n).

For the group of improved products we found no significant relationship between the involvement of technology-related actors and the products' short-term market performance in both stages of the innovation process. The long-term market performance of improved products is related significantly to the involvement of companies supplying ingredients and other inputs as source of innovation and as partner; however both in a negative way. None of the market-related actors play a significant role in the improved product's performance, either in the idea stage, or in the development stage.

5.5 Discussion and conclusions

The present study reveals how the openness and composition of product-related networks of external actors involved in the innovation process of new and improved products of the F&B industry relate to the products' short and long-term performance on the market. The results of our study support the hypothesis which stated that the more open the innovation network, the better the product's short-term and long-term market performance. This applies to openness of the networks both at the fuzzy front end – i.e. in the idea generation stage – and during the development stage of the product innovation process. More specifically we found that the use of more different external sources of innovation is a key factor for product innovations to become successful in the long term, for the whole sample and more specific for the group of new products (Section 5.5.2, Table 5.5). This can be explained by the following: companies that develop new products that are designed to stay for a long period on the market need to perform more extensive scanning at the fuzzy front end of the innovation process and check all possible sources. Products that are designed to stay for a shorter period of time – as they are regularly renewed – need less extensive searching and checking of sources.

Openness in terms of involving more different external actors in the development stage of the innovation process is significantly affects market performance for the better, especially short-term market performance of new products. Openness of the innovation network is not a crucial factor for successful short or long-term market performance of improved products.

We found that improved products in the F&B industry will show a better short-term market performance, while new products will show a better long-term market performance. This is not in line with what several authors found. An empirical study of Kleinschmidt and Cooper (1991) found a U-shaped relationship between success and degree of innovation. Similar results came from Van der Panne (2004) who observed that radical innovations tend towards high-risk-high-return patterns, showing sales records that are above but also below expectations. However, the literature on this issue remains inconclusive, as for instance Zirger (1997) – in line with our results – found a linear relationship between degree of innovativeness and the product's success. In our case we might explain the impact of the product's innovativeness on the product's long-term success from the fact that these products stand out in product advantages for the customer. We found a significant positive relationship between product innovativeness and long-term performance: the more innovative in terms of numbers of new product attributes, the better its long-term performance. Product quality – given the many new aspects – is highly appreciated by customers, which confirms what others found: a good quality product will always find a market (see also Link, 1987; Rource and Keeley, 1990; Calantone *et al.*, 1993). The high success rates of the products we found – nearly two thirds of the products that were introduced in the second half of 1998 as new products were still on the market seven years later, at the end of 2005 – might be explained by the fact that these are products that were selected by the editorial board of the professional journals. They could very well have chosen products which were worth being announced as they had a good chance of being successfully marketed. In literature overall very high failure rates (72-88%) are reported for food products that are introduced to the market (Buisson, 1995; Rudolph, 1995; Van Poppel, 1999; Lord, 2000). However, they usually fail in the first year after market launch; the products in our sample have passed this critical phase.

Our investigation of the impact of the involvement of technology-related and market-related actors in both stages of the product innovation process on the products' short-term and long-term market performance showed some very interesting outcomes. First of all we found that the composition of the innovation networks of new products differed from those of improved products. The first included relatively more technology-related actors as sources of ideas; the latter included relatively more market-related actors (in both stages of the innovation process) and technology-related actors in the product development stage (see Section 5.5.1, Table 5.3). However, the differences in the networks are very small, except for the involvement of a number of market-related actors (customers as partner, market intelligence providing companies as outsourcer) for which the networks differ significantly from each other. Although the level of involvement of external actors in the innovation networks of new versus improved products is rather similar; we found that their roles in the product's success differ considerably between the two product groups. This was also what we expected.

We found for new products that the involvement of both technology- and market-related actors contributes significantly positively to the product's market performance. The customer and competitor as source of ideas, the research organisations to which activities are outsourced and the companies that supply machinery/equipment contribute the most to long-term success. The companies supplying ingredients/raw materials and machinery/equipment as sources of ideas, the companies to which market-related activities are outsourced and the companies that provide training and that sell machinery/equipment contribute the most to the short-term success of new products.

Our finding that customers greatly affect long-term performance when they operate as source of innovation is in line with Maidigue and Zirger (1984) who found that the majority of successful ideas originate from the market and not from inside the company. However, we also found that the involvement of customers is not a factor for short or long-term success, although they are relatively well involved as partner in the development phase (see Section 5.5.1, Table 5.3). This could be explained as follows: as customers express their preferences in terms of already familiar products, customers bias innovators towards more incremental products. Involving customers as partners might diminish creativity and make the company disregard technology-driven ideas leading to more innovative products (Ortt and Schoormans, 1993; Wind and Mahajan, 1997; Bonner and Walker, 2004). Knudsen (2007) even found a significant negative relationship between involvement of customers in food innovation processes and innovative performance and argues that this might be due to the short-term orientation of the study. Our study also shows that in the long term involving customers as partner has neither a significant positive or negative impact on the product's success on the market. However, there is no agreement on involving customers in product development, as for instance Van der Panne (2004) found that collaboration with customers has a positive impact on the product's success since it significantly reduces the risk of overestimating demand.

For improved products we found that the involvement of companies that supply ingredients and raw materials as source of innovation and also those that supply machinery, equipment as partners of the innovating company strongly negatively affect long-term market performance. Although they are the most frequently involved external actor in the innovating company, they should not interfere with the innovation process: when they get too close this negatively affects the company's performance. Appropriability problems between the innovating company and its suppliers when the latter get too closely involved could be a reason for this negative relationship. Also, and this is an argument mentioned by Knudsen (2007), choosing partners from one's own industry (which is the case with suppliers of ingredients and raw materials) might imply a lack of cross-disciplinarity and thus missing conditions for new product ideas. However, a most convincing explanation could be the following. In the period since the end of the 1990's when the products were introduced on the market, many F&B companies have rationalised their purchasing activities. Up to that time, companies had many different ingredient suppliers which they invited to come with proposals for new ingredients for new or improved products they had in mind. However, at the beginning of this century companies changed their purchasing strategy and selected only a few with whom they could cover their whole product portfolio. In that case, companies have withdrawn products with ingredients from companies that did not belong to their core suppliers from the market. The case of Unilever, where the product development department worked very closely together with a small group of three suppliers of breakfast product ingredients, illustrates this change (Smit et al., 2008). This change in purchase strategy could very well explain our outcomes.

The contribution of the study lies in the focus on the product – Schumpeter's object of industrial renewal – and not on the innovating company. This enabled us to investigate the openness and composition of the product-related innovation network and the impact of actors in this network on the product's market success. A second important contribution of this study is that the product's market performance was not only measured just after market introduction, as is usually the case, but also for a second time seven years after its introduction to the market. This enabled us to investigate which actors in the innovation network are decisive for products that provide companies with income for many years, compared to those that provide short-term gains. Most important was was the discovery of significant differences in the openness of the product-related network that affect short-term versus long-term success of new products versus improved products. The involvement of technology-related (research institutes, companies providing training, companies supplying machinery and equipment) and market-related (customers, competitors, companies providing market intelligence) proved to have a significantly positive impact on both short-term and long-term market performance of new products, but not on that of improved products.

Our findings have important managerial implications for innovating companies, when it concerns the selection and role of external actors in the idea generation and product development stage of the innovation process. They have to select their partners carefully. The best cooperation strategy with technology-related knowledge institutions is to outsource to research institutes, rather than to universities or polytechnics. Research institutes (as defined in our study) focus on application-oriented research; this is much closer to what industry needs, as compared to universities which focus on basic research. The polytechnics' role in contract research is still very minor.

Our study shows that companies supplying technology-embodied goods and services are important actors for the innovating companies as they are frequently involved in the product's innovation process. However, their role has to be very carefully defined, as we found that as source of innovation or as partner in the improved product's innovation process they negatively impact on the product's success. Cooperation with those companies that bring new technological developments that have been developed elsewhere embedded in the provided goods and services in their role as seller seems to be the best cooperation strategy for innovating companies.

Market research and market orientation are critical activities when bringing new F&B products successfully to the market (see for instance Traill and Grunert, 1997; Omta *et al.*, 2003; Grunert *et al.*, 2008). On the basis of our results we recommend that companies planning to develop new products seek out a large variety of sources of ideas for innovations, especially among customers. However, companies bringing improved products to the market are recommended not to spend too much time involving any external market-related actor as we found that none of them had a specific impact on the product's success. In this case the company's internal market-related competences could be sufficient for successful market introduction.

6. Discussion and conclusions

In this chapter we discuss the main findings of the different studies presented in this book. To do so, we elaborate on the results of the research questions and hypotheses in Section 6.1. Section 6.2 draws more general conclusions and addresses the main research question of this book. In Section 6.3 we summarise the main contribution to literature provided in this book. Finally, this chapter closes in Section 6.4 with suggestions for further research.

The main research question to be answered in this book is:

> *What key factors are positively related to the short- and long-term market performance of F&B products?*

For our empirical investigation we have built a database of products that have reached market introduction. This discards innovations that fail prior to market introduction and allowed us to study the impact of various factors on the market performance of the products. We collected data on newly launched products that were announced in Dutch trade journals in the second half of 1998. The data were collected in 2000 through interviews with managers that were directly involved in the product development process. In December 2005 additional data on the long-term market performance of the products were collected. The total sample of 129 products included 76 products for the consumers market, 37 for the industrial market and 16 for the food services market. The products had been produced by 66 different Dutch F&B companies.

We found that 64% of the products in our sample were still on the market seven years after product launch. In the literature overall high failure rates are reported for F&B products that are introduced to the market (Buisson, 1995; Rudolph, 1995; Lord, 2000). The products in our sample were on the market one and a half years after market launch, which means that they had passed the critical first year, where about one out of three newly launched F&B products for the retail market fail (Van Poppel, 1999). This is one explanation for the unexpected high success rate. The method we have used for the identification of the products might also provide some explanation. We used products that were selected by the editorial board of the professional journals. It can be assumed that this board chose those products which they assumed had a good chance of being successful in the market.

6.1 Main findings

The study in Chapter 2 focuses on the internal organisation and management of the innovation process. It investigated how the company's in-house resources are successfully managed, with a specific focus on the role of product innovativeness. The research question is:

RQ1: *Which technology- and market-related resources play a key role in the short- and long-term market performance of new versus improved products?*

We have developed a model that integrates three key aspects that are fundamental to a company's success in innovation: upfront activities, organisational routines and company culture. Upfront activities dealing with the execution of predevelopment activities are rewarded in reduced development time and improved market performance and profitability (Cooper and Kleinschmidt, 1987; 1990; Cooper, 1993). We have distinguished between activities that involved technology-related and market-related in-house resources. Organisational routines are the essence of the firm's dynamic capabilities: they form the particular behaviour of a company that has emerged as a result of doing things over a period of time and what has appeared to work well (Teece *et al.*, 1997). The company culture is the pattern of shared values, beliefs and agreed norms which shape behaviour. It can raise the awareness, creative skills and problem-solving abilities of all employees and can lead to higher levels of participation in innovation (Tidd *et al.*, 2005).

As there is no consensus in the literature on what is most important in product innovation of new versus improved products – technology-related resources or market-related resources – it was decided not to formulate hypotheses; for that reason the study had an explorative character.

We found that the management of the innovation process of new products differs significantly from that of improved products. This applies to the involvement of technology-related (technical feasibility) and market-related (preparation and testing of the product by consumer) resources as part of upfront activities, the involvement of the company's R&D functions and the use of specific organisational routines (product development plans, ex-post evaluation). They also differ in the scale of preparatory and test activities and the number of organisational routines. In all cases these activities and routines were more often applied in the development process of new products. Apparently the development of new products demands a more organised and structured process than that of improved products, including both technology-related and market-related resources.

In addition, it was found that new products not only performed better in the long term, but also in the short term. Apparently, the market is able to recognise and appreciate the qualities of new products. Our findings support the statement that the vision of the consumer as having a predominantly conservative and inert attitude towards new developments in food products and even as having a specific risk aversion in their choice (Galizzi and Venturini, 1996) needs to be revisited.

In answer to research question 1: for new product's successful short-term market performance a number of both technology- and market-related in-house resources (activities, company functions involved) are critical. The technology-related resources included: technical feasibility, test of the production process, and the involvement of the company's production function. The market-related resources included marketing research, test of product concept and of product by customer and the involvement of the company's marketing function. For improved products, only one market-related actor is of significant importance: the involvement of the company's sales function will do. Technology-related resources are much more critical for the improved products' short-term market performance. They include the company's R&D function both as source of ideas and its involvement in the product development process, technical feasibility and test of the production process. With respect to long-term market performance it was found that a combination of technology-related resources (R&D function involved and test of product quality) and market-related resources (marketing research and test of product concept with customers) are only crucial for improved products; none for new products. We concluded that – also due to the many major uncertainties that accompany the development of new products –, other innovation trajectories and management systems can play a key role in the long-term market performance of these products. See Table 6.2 for an overview of the in-house technology and market-related resources for new and improved products (2nd and 4th columns).

With respect to organisational routines there are some differences between the two product groups: for new products the use of product development plans and for improved products making no/no-go decisions are crucial for short-term market performance. For both products groups a prepared product development process is a factor for successful short-term market performance. Interesting differences were found for company culture: a company culture that favours cross-functional cooperation (especially between technology-related and market-related functions) and that is focused on results is a prerequisite for a new product's short-term market performance. Surprisingly, it was found that not for new product but for improved products to stay successful on the market, a company culture that is characterised by flexibility and openness is of crucial importance. The latter cultural characteristics are often associated with innovative products (see also Earle *et al.*, 2001) as they deal with conditions that are required for companies that want to stimulate creativity, capitalise on new chances, give room to individual initiatives and facilitate the trial-and-error character of the new product development process (Peters and Waterman, 1982). However, our study shows that these are also crucial for companies that want to pursue a high level of continuous innovation, specifically the constant improvement of their core products.

Unexpectedly, none of the resources and other factors studied showed a significant impact on the long-term market performance of new products. Apparently, there are many different innovation trajectories and management system that lead to these products' long-term market success. Overall, our results suggest that it is possible to manage product innovation for short term, but not for long-term market performance.

Chapter 3 also focused on the innovation process. Here we explored the differences in the use of both in-house and external resources in the innovation processes of products for the consumer and the industrial markets. In this chapter we addressed research question 2:

RQ2: *What are the differences in the involvement of technology-related and market-related resources in the innovation process of consumer versus industrial products and in their short- and long-term market performance?*

More specifically, we tested the two following hypotheses:

H2a: *The innovation process of consumer products will include more market-related resources, whereas that of industrial products will include more technology-related-resources.*

H2b: *Consumer products will show a better short-term market performance, whereas industrial products will show a better long-term market performance.*

We found only limited support for Hypothesis 2a. Only in the development stage of the consumer product innovation process two market-related resources (more specifically: the company's marketing function and outsourced marketing-related activities) were involved significantly more often compared to industrial products. In the idea stage market-related sources of innovation were used more often in the case of industrial products (this applied to customers and competitors). For technology-related resources we found that in the idea stage they were used more often in the innovation process of industrial products (patent literature). However, in the development stage technology-related resources (suppliers of ingredients and or equipment) were used significantly more often in the case of consumer products. We therefore conclude that Hypothesis 2a cannot be confirmed. Both market-related and technology-related resources were involved significantly often in the innovation process of both product groups.

For consumer products it was shown that suppliers in particular were involved in the innovation process significantly more often than for industrial products. It could very well be that these suppliers were producers of ingredients. According to Joppen (2004), companies that supply ingredients, nowadays also provide total product concepts that include both ingredients and recipes for producing the consumer products. For that reason their involvement not only represents technology-related resources, but they also fill an important information gap by providing information on developments on consumer markets. This is certainly (part of) the explanation for the important role of market-related resources in the innovation process of industrial products.

Hypothesis 2b could be confirmed for industrial products as they showed an overall better performance in the long term than consumer products. However, also in the short term they performed better than consumer products. So consumer products not only stay for a shorter period of time on the market, but also contribute less to a company's turnover and market share, compared to industrial products. The better performance of industrial products, i.e. the ingredients, could be explained by the multiple uses of ingredients in various consumer products and the fact that some of them can have long life cycles.

Chapter 4 focused on the company's product innovation strategy and how this relates to the product's market performance. The research question that was addressed in this chapter is:

RQ3: *What is the impact of the product innovation strategy on the product's short- and long-term market performance?*

A company's innovation strategy forms an essential element of its overall business strategy. It provides guidelines for dealing with strategic questions, such as which new products to develop, for what markets, which necessary expertise to develop within the company and which expertise to acquire from outside. Miles and Snow (1978) have developed a theoretical framework of strategic archetypes (prospector, analyser and defender), each showing a stable and consistent pattern of response to the changing environments in which the companies that follow such a strategy operate. Using this theoretical framework and the argumentation that developing new products demands considerable investments that can only be earned back if these products stay on the market for a long period of time, we formulated the following hypothesis:

H3: *The products of F&B companies that intend to follow a prospector innovation strategy will show better long-term market performance than the products of companies that intend to follow another (analyser, defender) innovation strategy.*

Our findings showed that, unexpectedly, companies that focused on a prospector strategy in combination with performing extensive market assessment activities in preparing the product development process, are successful in the short term, but not in the long term. Pursuing a defender strategy leads to long-term market performance. Companies following the prospector strategy maintain a high level of innovative activity by regularly introducing new products to the market, thereby replacing their 'old new products'. On the contrary, once companies following a defender strategy have introduced a new product, they are less likely to engage again in developing new products, to replace the older ones. Instead, they will try to keep their products on the market for as long as possible. Companies following a defender strategy maintain – by regular small improvements – and also protect their once introduced product on the market on a certain level of successful market performance. The results suggest that a more segmented categorisation of company strategies is more appropriate for understanding what strategic factors influence the products' successful market performance.

In Chapter 5 the impact of the innovation network on the product's short- and long-term market performance was investigated. In this chapter we addressed the following research question:

RQ4: *What is the impact of the openness and composition of the product-related innovation network on the short- and long-term market performance of the product?*

Openness refers to 'open innovation' (Chesbrough, 2003) which stands for the increase in the use of external resources by innovative companies in order to speed up the innovation process. F&B companies rely more on external sources of innovation than the average for all industries: the interactions between companies and their business partners in the supply chain as well as with public research organisations play a crucial role in achieving successful innovations. Most F&B companies have only limited in-house resources for scientific and technological activities; as a consequence they will have to develop a broad interface with actors that provide them with scientific and technological input (Christensen *et al.*, 1996; Rama, 1996; Knudsen, 2007). Also F&B companies need extensive market knowledge in order to have a better view of long-term consumer needs. For this they use input from different market-related actors that provide them with the necessary market intelligence (Grunert *et al.*, 1996, 1997, 2008; Sarkar and Costa, 2008). We expected that F&B companies that innovate in networks are more successful than companies that don't. Although there will be a limit to the number of external relationships that can be managed by a company successfully, we expected that this restriction would not apply to F&B companies. The following hypothesis was used:

H4: *The more open the product-related innovation network, the better the product's short- and long-term market performance.*

This hypothesis was confirmed and applied to both the idea stage and the development stage of the product innovation process. When investigating the role of product innovativeness, it was found that the relationship between openness of the product-related innovation network and the product's market performance was significant for both new and improved products, but most significantly for the new products group.

We expected to find different levels of involvement of external technology- and market-related actors in the product's network and also differences in the impact of these resources on the product's short and long-term market performance, depending on the innovativeness of the product. For new products we found that both technology- and market-related actors contributed significantly positively to the product's market performance. These actors can be involved as source of innovation in the idea stage and as partner, outsourcer of seller in the development stage of the product innovation process. The customer and competitor as market-related source of ideas, the research organisations to which activities are outsourced and the companies that supply machinery and/or equipment (both technology-related) contributed most significantly to long-term market performance. Most important for the short-term market performance of new products were the companies that supplied ingredients and/or raw materials and machinery and/or equipment who provided sources of ideas for the new products, the contribution of companies to which market-related activities has been are outsourced and the companies that provided training and that sold machinery and/or equipment.

For the group of improved products we found that the involvement of technology-related actors, and more specifically companies supplying ingredients and/or raw materials as source of innovation and companies that supplied machinery and/or equipment and those that supplied raw materials and/or ingredients as partner of the innovating company strongly negatively affected the long-term market performance of these products. A most probable explanation could be that the relationship with many ingredient and raw material suppliers was stopped – due to the rationalisation of suppliers at the beginning of the 2000s – and the products with these ingredients had been withdrawn from the market. Appropriability problems between the innovating company and its suppliers and a lack of cross-disciplinarity and thus missing conditions for product ideas when linking with partners from one's own industry, could also explain this outcome. See Table 6.2 for a summary of the results: it presents the technology- and market-related actors in the product-related innovation networks of new and improved products that have significantly contributed to the product's short and long-term market performance (3rd and 5th column).

To answer research question 4, we refer to the above outcomes which provide insights into the impact of the openness and composition of the product-related innovation network on the short and long-term market performance of new versus improved products.

Table 6.1 provides an overview of the outcomes of the empirical tests of the hypotheses of the studies presented in Chapters 3 to 5. The study in Chapter 2 had an explorative character; no hypotheses were used.

6.2 Discussion and conclusions

Combining the findings of the four studies enables us to draw overall conclusions and to provide an answer to the central research question:

RQ: *What key factors are positively related to the short- and long-term market performance of F&B products?*

The main focus of our study is on the factors that drive and enable product innovation in the F&B industry. This industry is characterised as a low- and medium-tech industry. According to Pavitt (1984) low- and medium-tech industries are mainly characterised by process innovations. Also firms in the F&B industry were found to be more process-innovation oriented (Grunert

Table 6.1. Outcomes of the empirical tests of the hypotheses.

Hypotheses	Outcomes
Innovation process: consumer versus industrial products	
H2a: The innovation process of consumer products will include more market-related resources,	not confirmed
whereas that of industrial products will include more technology-related- resources.	not confirmed
H2b: Consumer products will show a better short-term market performance,	rejected
whereas industrial products will show a better long-term market performance.	confirmed
Innovation strategy: prospector versus defender and analyser strategies	
H3: The products of F&B companies that intend to follow a prospector innovation strategy will show a higher long-term market performance than the products of companies that intend to follow another (analyser, defender) innovation strategy.	rejected
Innovation network: open innovation	
H4: The more open the product-related innovation network, the better the product's short and long-term market performance.	confirmed

et al., 1997). Christensen *et al.* (1996) qualified the F&B industry as a 'carrier' industry: an industry that puts into widespread use the new technologies developed in high-tech industries.

There is a growing focus on the importance of the market-driven character of innovations in the F&B industry and especially on the role of users, i.e. consumers (see for instance Galazzi and Venturini, 1996; Christensen *et al.*, 1996; Grunert *et al.*, 1996, 2008). Moreover, there is a growing body of evidence that advocates putting the consumer at the start of the 'food chain' (Urban and Hauser, 1993; Saguy and Moscowitz, 1999; Lord, 2000).

Market-orientation and consulting customers as sources of information is crucial for the innovation process; incorporating their interest is a factor for successful innovation (see for instance Cooper and Kleinschmidt, 1987; Montoya-Weiss and Calantone, 1994). This applies also to the F&B industry: market-driven food product development processes increases the success ratio for new products considerably (Grunert *et al.*, 1996, 1997; Hoban, 1998, Kristensen *et al.*, 1998; Knox *et al.*, 2001; Stewart-Knox and Mitchell, 2003). This seems a logical consequence of the recent developments in agrifood supply chains where the market has become a major driver for innovation processes. This could suggest that technology plays a minor role in product innovation in this industry. However, the technological aspect of innovation in this industry might be underestimated (Traill and Meulenberg, 2002).

Table 6.2 provides an overview of the technology- and market-related internal and external resources that were involved in the innovation processes of new and improved products and that proved to be significantly related to short and long-term market performance (based on the findings in the Chapters 2 and 5).

When analysing the importance of internal technology- and market-related resources for the market performance of new products, we found that both are crucial for their successful short-term market performance. For the improvements of products, the use of technology-related resources in the idea stage and in the preparatory and test stage of the product innovation process were crucial for short-term market performance; whereas market-related resources did not play a significant role here. These technology-related resources were involved in order to improve the product's characteristics which are a determinant for successful continuous innovation (Van der Panne, 2004). Moreover, for sustained market performance both technology-related and market-related upfront activities were needed: the R&D function needs to be involved, but market-related activities (marketing research, testing by customers) also turned out to be important for the product's long-term market performance.

However, especially the findings on the open character of the product innovation process and the composition of the external innovation networks showed that technology-related resources were most influential. The external actors that have an impact on the new and improved product's market performance are mostly technology-related actors, especially in the development stage of the product innovation process. Also the findings in the study on consumer versus industrial products showed that both factors are important for the product's success on the market (Sections 3.4 and 3.5).

Our findings show that, although the F&B industry is characterised as a medium- to low-tech industry according to formal statistics based on in-house R&D activities and the focus in literature is mostly on the market-driven and user-oriented character of innovations in the F&B industry, in the product innovation process technology-related resources were very important. This importance of technology might increase in the future as F&B companies are increasingly confronted with competition from private label products and – in order to improve their competitiveness – they choose to become more innovative, preferably in products and processes that have proprietary elements that can be protected. Together with the argument that R&D may be underestimated in the F&B industry because of the lower degree of functional specialisation in small and medium-sized firms (Kleinknecht *et al.*, 1991) and the trend for healthy ingredients to become more and more knowledge intensive, this might explain the increasingly technology-based character of product innovations in the F&B industry. In the literature a revision of the classification of industries that is solely based on

Table 6.2. Correlation between technology-related and market-related resources and short- and long-term market performance of new and improved products.

| | **Technology-related resources** | |
	Internal [1]	External [2]
New products		
Short-term market performance	Company function involved: • production * Preparatory and test activities: • technical feasibility *** • test production process ***	Source of ideas: • equipment companies * Seller: • training *** • equipment companies **
Long-term market performance		Outsourcer: • research organisations ** Seller: • equipment companies *
Improved products		
Short-term market performance	Source of ideas: • R&D department ** Company function involved: • R&D ** Preparatory and test activities: • technical feasibility ** • test production process *	
Long-term market performance	Company function involved: • R&D ** Preparatory and test activities: • product quality –*	Source of ideas: • ingredient companies –** Partner: • ingredient and equipment companies –**

Significant correlations are indicated as follows: *** for $P<0.01$; ** for $P<0.05$ and * for $P<0.10$; – = negative relationship.

[1] Internal resources refer to the in-house sources of ideas, company functions involved and preparatory and test activities applied during the product development process.

[2] External sources refer to the actors in the product-related innovation network. They can be involved in four different roles: source of ideas, partner, outsourcer or seller (See Table 5.1 for definitions).

Market-related resources

Internal

External

Company function involved:
- marketing **

Preparatory and test activities:
- marketing research **
- test product concept by potential customers **
- test product by customer *

Outsourcer:
- marketing companies **

Source of ideas:
- customers ***
- competitors **

Company function involved:
- sales *

Preparatory and test activities:
- test product concept by potential clients **
- marketing research *

R&D-intensity is being discussed (Hirsch-Kreinsen *et al.*, 2005; Peneder, 2007; Reinstaller and Unterlass, 2008).

A second conclusion deals with the role of suppliers. Low- and medium-tech industries rely heavily on embodied technologies for improved productivity; they correspond largely to Pavitt's (1984) class of 'scale-intensive' and 'supplier-dominated' industries. Our results show that ingredient suppliers more than suppliers of equipment (16% versus 5%) were used as sources of innovation, suggesting that suppliers are more involved in product innovations than in (the related) process innovations. Suppliers of ingredients were sources of innovation for both consumer and industrial products and on a rather equal level: 16% for industrial products and 17% for consumer products. Earle *et al.* (2001) argue that food ingredient suppliers have gone a step further than just having a good relationship with the F&B manufacturer; they often develop the related manufacturing processes. Our results also show that suppliers play a significant role in the product innovation process. Next to their more passive role as seller of their products (29% of the cases), we found that in more than one third of the cases (35%) they were involved as partner in the product development process. This suggests that F&B firms actively operate in alliances with their suppliers to improve their products and processes. However, we also found that the involvement of suppliers has a negative impact on the products market performance. The trend in the beginning of the 2000s when F&B companies rationalised the large numbers of ingredient suppliers and started to work only and much more closely with a small number of preferred suppliers could explain this outcome. This finding poses the interesting question as to what is the role of supplier involvement in innovations in the F&B industry. On the basis of our findings, we suggest that the vision of the role of suppliers in innovation in the F&B industry has to be revisited; new types of suppliers (providing ingredients, instead of equipment) are now contributing significantly to product innovation in this industry and their role in the innovation process is changing, illustrating the opening up of the innovation processes. So the dynamics in the involvement of suppliers in F&B innovation processes has to be studied carefully.

A third conclusion deals with the role of customers and the mechanism by which information on market developments and other relevant market-related input is channelled into the product innovation process. We found in Chapter 3 that customers (i.e. producers of consumer products) were mostly involved in the innovation process of industrial products. Unexpectedly, we found hardly any involvement of customers in the innovation process of consumer products, although many sources – mostly focussing on consumer products – underline the role of market-pull in innovation processes and of involving customers (retailers, consumers) in product development processes (including Galazzi and Venturini, 1996; Christensen *et al.*, 1996; Grunert *et al.*, 1996; Dahan and Hauser, 2001). We expected that the closer to the market the more often consumers directly or indirectly (through retailers) would play an important role in the product innovation process. This is not what we found. We found that only the company's marketing function played a significant role in this respect. Market intelligence in this sector is essential for creating long-term competitive advantage, but it is still relatively poorly developed (Costa, 2003). The absence of relationships with customers could imply that market-related information has come through other channels. On the basis of another finding of our study which showed the strong relationship of producers of consumer products with their ingredient suppliers (see Table 3.4), we concluded that this could represent a channel through which market knowledge is being transferred. So market intelligence could be gathered by consumer product companies from a number of different channels; not only through interactions (of the company's marketing and sales function) with retailers or consumers directly and but also indirectly for instance through their ingredient suppliers.

In the study on innovation networks of new versus improved products we found that customers were involved as partner in the development phase of both product groups (39% versus 28%), but their involvement as partner proved not to be a factor significantly positively related to short- or long-term market performance of these products. This finding confirms results of other studies which found that too close an involvement of customers might diminish creativity and make the company to disregard technology-driven ideas leading to more innovative products; customers express their preferences in terms of already familiar products and thus bias innovators towards more incremental innovations (Ortt and Schoormans, 1993; Wind and Mahajan, 1997; Bonner and Walker, 2004). Knudsen (2007) even found a significant negative relationship between the involvement of customers in food innovation processes and innovative performance; she argues that this might be due to the short-term window of 1-2 years of her study. Our study measured market performance after seven years, and showed that involving customers as partner in product development had no significant impact on the product's market performance after that period.

Although the level of involvement of external actors in the innovation networks of new versus improved products was rather similar, we found that openness of the network had a clear positive impact on the new product's short-term and long-term market performance. Apparently, the inputs from external resources in the conceptualisation and development stage of new products are very essential and give characteristics to the product that are also valued by the market more than for improved products.

F&B companies increasingly operate in a competitive environment in which changes in competitor strategy, new technological opportunities, changes in market trends and new governmental regulations can make the difference between success or failure. Those companies that stay alert to what is happening in their environment and that closely assess potential future developments are able to keep ahead of their competitors. Therefore the most critical questions to be answered by the management of innovating companies are: what products do we want to make and for which markets (strategy), which partners and internal functions do we need to incorporate (resources) and how can we facilitate and coordinate the innovation process (implementation)? It is essential for companies achieving success in innovation to sustain their market position and keep ahead of their competitors. From a managerial perspective it is therefore important to know how to improve the innovation process. Based on our results in this book we can conclude that it is important for the product's long-term success in the market that companies have a clear orientation that determines their short- and long-term product innovation strategies. Such a strategy should hold a combination of strategic considerations for the short term based on prospector type of strategies in combination with extensive market assessment activities and for the long term based on defender type of strategies. Moreover, for sustainable success a company culture that is characterised by flexibility and openness is of significant importance. A culture that stimulates creativity, provides room for individual initiatives and facilitates the trial-and-error character of the product development process is crucial for companies that want to achieve a high level of continuous innovation, specifically the constant improvement of their core products.

6.3 Contributions to literature

In this book we present empirical findings regarding the key factors for successful product innovation in the F&B industry. Our study has provided the following main contributions to literature.

First, our study reveals some key insights into the role of technology-related resources and market-related resources in successful product innovation processes in low- and medium-tech industries. Our empirical evidence strongly supports the view that non-R&D resources are crucial in understanding the innovation process in this industry, the studies in this book have shown in particular that using customers as source of innovation, performing technology assessment, testing of production and training are decisive factors for innovation outputs. Companies in low- and medium-tech industries have alternative sources beyond R&D that may be highly useful for achieving successful product innovations. Although the company's R&D function is an important source of innovation and also plays a crucial role in the product development process, our empirical evidence confirms the importance of other sources and resources from inside and outside the company.

Second, our study focuses on the innovative product, rather than on the company: Schumpeter's object of industrial renewal. There is growing recognition of the role of the level of product innovativeness in the product's market performance (Hoban, 1998; ECR Europe, 1999; Lord, 2000). However, there is a lack of studies providing insight into which factors dealing with product innovation process, strategy and network are positively or negatively related to the short and long-term market performance of new versus improved products. Our study is the first that provides empirical evidence of these relationships.

Third, we were able to analyse the market performance of a specific product not only soon after market launch, but also after several years (i.e. the product's short and long-term market performance). Basically for practical reasons, there is hardly any research being done so far on the long-term performance of new and improved products. This study aims to fill that gap. Moreover, most studies that analysed key factors for successful innovation in the F&B

industry used more indirect performance measures such as expected turnover growth (Traill and Meulenberg, 2002) or failure of success as defined by the company (Stewart-Knox *et al.*, 2003), or deal with input indicators, such as R&D expenditures and human resources (Avermaete *et al.*, 2004). We used measures for commercial performance of the products on the market.

Fourth, our research is one of the first studies on open innovation in the F&B industry. Empirical investigations on open innovation in the F&B industry are scarce (Sarkar and Costa, 2008) although there is a growing number of companies in this industry that use open innovation strategies (Huston and Sakkab, 2006; Vanhaverbeke and Cloodt, 2006; Smit *et al.*, 2008). However, open innovation was perceived to be especially important for large and high-tech firms (Chesbrough, 2003). Our empirical study quantified the relevance of open innovation for low and medium-tech sectors such as the F&B industry. As such our empirical findings can be regarded as an important contribution to the literature on open innovation.

Fifth, our study is one of the first to carry out an in-depth investigation into the technology-related aspects of product innovation processes in the F&B industry. There still is a lack of studies on the role of technology in product innovation in this industry (Rama, 1998, Wilkinson, 1998, Traill and Meulenberg, 2002). The low R&D intensity of the sector might also have influenced the lack of focus in innovation studies on technology-related aspects. This can partly be explained by the dominant user-orientation in F&B innovation studies (see for instance Grunert *et al*, 2008), but also because technology-related factors do not fit in with the rural, authentic and non-industrial image F&B companies want to associate their products with. In addition, the adverse reactions of specific consumer groups to some technologies (radiation, GMO food) might have drawn the attention away from technology. This study aims to fill this gap and has investigated both technology and market-related factors. By doing so, our study provided relevant insights into the ongoing discussion on the importance of technology versus market factors in innovation processes in low- and medium-tech industries.

Last, we want to note the potential for generalisation of the findings of this study. Although our data are limited to the Netherlands, we would argue that generalisation is permissible for countries that also have low-tech industries with a relatively high performance, as the Dutch F&B industry. As low- and medium-tech industries are often major customers of high-tech innovators (Robertson *et al.*, 2003), together they form the essential pillar in advanced regional or national economies. For that reason we can conclude that this generalisation applies to countries that have a combination of relatively high performing low-, medium- and high-tech industries and would include most Scandinavian and other North-western European countries (Robertson and Patel, 2007).

6.4 Suggestions for further research

The results of this study suggest the need for further research.

The first suggestion relates to the importance of technology-related versus market-related resources in product innovation processes. The results of our study require a more detailed investigation of how and through which mechanisms the involvement of technology-related resources influences product innovation in the F&B industry. In our study we have used a number of proxies for technology-related factors related to strategy, open innovation and management of product development processes. However, these proxies are only crude representations of the underlying concept of the technology-relatedness of each of these aspects. Traill and Meulenberg (2002) argued that the marketing literature has developed a number of scales to measure various aspects of the market orientation concept. Similarly it is recommended to focus future research in the field of product innovation processes in low- and medium-tech industries on developing an analytical framework dealing with technology-orientation. Such a framework for technology-related concepts should include variables and scales, for instance for measuring company competence levels associated with this orientation and other competencies that might be important for the product innovation process, such as for interaction with external scientific and technical competences. This could include research on the importance of the various channels through which directly (in-house and external actors) and indirectly (literature, patents, products and services) technological knowledge flows into the product innovation process. This would contribute to a more complete analytical framework of factors driving product innovation in low- and medium-tech industries.

Based on our conclusions regarding the role of external actors in open innovation in the F&B industry, we suggest that further research is needed in order to fully understand how the involvement of suppliers, customers and research organisations contribute to innovation performance in low and medium-tech industries, such as the F&B industry. There is, mainly in marketing literature, considerable attention devoted to the role of the market and more specifically the user as driver of innovation processes (e.g. Grunert *et al.*, 1996, 2008), while technology-related determinants which are expected to play a modest role due to the 'low tech fix', have virtually disappeared (see also Traill and Meulenberg, 2002).

Firstly, an attempt to include technology-related determinants would contribute to a new area on the research agenda on user-oriented innovation in the F&B industry (Grunert *et al.*, 2008). This agenda covers a range of research issues, including questions on the phenomenon itself, about the barriers towards user-oriented innovation, the optimum business models and related innovation strategies and that of the interdependency of bringing new products to the market and the development of users needs (*ibid.*). We propose that the new research area would deal with the role of technology-related resources in user-oriented innovations as well. Research in this area should particularly focus on questions such as: How technology plays a role in consumer choice on the one hand and how technology can help to develop products that meet the developing consumer preferences concerning food quality and safety on the other.

Secondly, it would contribute to a more in-depth knowledge of in which type of processes and through which actors involved in these processes, companies acquire knowledge of the (future) market. We concluded that market intelligence could be gathered from a number of different channels; through interactions (of the company's marketing and sales function) with retailers or consumers directly, but also indirectly, for instance through their ingredient suppliers. We also concluded that companies planning to develop new products should seek out a large variety of sources of ideas for innovations; especially customers and competitors are of particular importance. However, companies bringing improved products to the market should not spend too much time involving any external market-related actors as none of them had a specific impact on the product's success. Further study could investigate what role companies that produce industrial products and other external actors within and outside the supply chain play in gathering and channelling market-related information in low- and medium-tech industries.

References

Acs, Z.J. and D.B. Audretsch, 1988a. Innovation and firm size in manufacturing. Technovation 7: 197-210.

Acs, Z.J. and D.B. Audretsch, 1988b. Innovation in Large and Small Firms: An Empirical Analysis. American Economic Review 78: 678-690.

Alfranca, O., Rama, R. and N. von Tunzelmann, 2001. Cumulative innovation in food and beverage multinationals. Business & Economics Society International. Global Business & Economics Review - Anthology 2000: 446-459.

Anderson, A.M., 2008. A framework for NPD management: doing the right things, doing them right, and measuring the results. Trends in Food Science and Technology 19 (11): 553-561.

Archibugi, D., Cesaratto, S. and G. Sirili, 1991. Sources of innovative activities and industrial organization in Italy. Research Policy 20: 299-313.

Archibugi, D., 1992. Patenting as an indicator of technological innovation: a review. Science and Public Policy 19: 357-368.

Arundel, A., Van de Paal, G. and L. Soete, 1995. Innovation strategies of Europe's largest industrial firms. Results of the PACE survey. European Commission, DG XIII, Luxembourg.

Avermaete, T., Viaene, J., Morgan, E.J. and N. Crawford, 2003. Determinants of innovation in small food firms. European Journal of Innovation Management 6: 8-17.

Avermaete, T., Viaene, J., Morgan, E.J., Pitts, E., Crawford, N. and D. Mahon, 2004. Determinants of product and process innovation in small food and manufacturing firms. Trends in Food Science and Technology 15: 474-483.

Balanchandra, R. and J. Friar, 1997. Factors for success in R&D and new product innovation: a contextual framework. IEEE Transactions on Engineering Management 44 (3): 276-287.

Balbontin, A., Yazdani, B., Cooper, R. and W.E. Souder, 1999. New product development success factors in American and British firms. International Journal of Technology Management 17: 259-280.

Batterink, M.H., Wubben, E.F.M. and S.W.F. Omta, 2006. Factors related to innovative output in the Dutch agrifood industry. Journal on Chain and Network Science 6: 31-45.

Beckeman, M. and C. Skjöldebrand, 2007. Cluster/networks promote food innovations. Journal of Food Engineering 79: 1418-1425.

Bhargava, M., Dubelaaar, C. and S. Ramaswami, 1994. Reconciling diverse measures of performance. A conceptual framework and test of methodology. Journal of Business Research 31: 235-246.

Bonner, J.M. and O.C. Walker, 2004. Selecting influential Business-to-business customers in new product development: relational embeddedness and knowledge heterogeneity considerations. Journal of Product Innovation Management 21: 155-169.

Boon, A., 2001. Vertical coordination of interdependent innovations in agri-food industry. PhD Thesis. Copenhagen Business School, Copenhagen.

Booz, Allen and Hamilton, 1982. New product management for the 1980s, Booz, Allen and Hamilton Inc, New York.

Breschi, S. and F. Malerba, 1997. Sectoral Innovation Systems: Technological Regimes, Schumpeterian Dynamics, and Spatial Boundaries. In: Edquist, C. (ed.), Systems of Innovation: Technologies, Institutions and organizations. Pinter, London.

References

Brown, S.L. and K.M. Eisenhardt, 1995. Product development: past research, recent findings, and future directions. Academy of Management Review 20 (2): 343-376.

Buisson, D., 1995. Developing new products for the consumer. In: D.W. Marshall (ed.) Food choice and the consumer. Chapman & Hall, Cambridge, pp.182-215.

Burgelman, R.A., Christensen, C.M. and S.C. Wheelwright, 2009. Strategic Management of Technology and Innovation (5 ed.). McGraw-Hill, New York

Calantone, R.J., Benedetto, C.A. and R. Divine, 1993. Organizational, technical and marketing antecedents for successful new product development. R&D Management 23: 337-349.

Calantone, R.J., Benedetto, C.A. and S. Bhoovaraghavan, 1994. Examining the relationship between degree of innovation and new product success. Journal of Business Research 30: 143-148.

Calantone, R.J., Cavusgil, S.T. and Y. Zhao, 2002. Learning organization, firm innovation capacity and firm performance. Industrial Marketing Management 31: 515-524.

Carlsson, B., Jacobsson, S., Holmén, M. and A. Rickne, 2002. Innovation systems: analytical and methodological issues. Research Policy 31: 233–245.

Cassiman, B. and R. Veugelers, 2006. In search of complementarity in innovation strategy: internal R&D and external knowledge acquisition. Management Science 52: 62-82.

Chabral, J.E.O. and W.B. Traill, 2001. Determinants of a firms likelihood to innovate and intensity of innovation in the Brazilian food industry. Journal on Chain and Network Science 1: 33-42.

Chakravarty, B.S., 1986. Measuring strategic performance. Strategic Management Journal 7: 437-458.

Chesbrough, H.W., 2003. Open Innovation. The new imperative for creating and profiting from technology. Harvard Business School Press. Cambridge MA.

Chesbrough, H., 2006. Open innovation: a new paradigm for understanding industrial innovation. In: Chesbrough, H., Vanhaverbeke, W., West, J. (eds.). Open innovation: researching a new paradigm. Oxford University Press, New York.

Chesbrough, H.W. and A.K. Crowther, 2006. Beyond high tech: early adopters of open innovation in other industries. R&D Management 36: 229-236.

Christensen, J.L., Rama, R. and N. Von Tunzelmann, 1996. Industry Studies of Innovation, Using CIS data. Study on Innovation in the European Food Products and Beverage Industry. Report for the European Commission. EIMS SPRINT, Brussels.

Christensen, C., 1997. The Innovators's Dilemma. Harvard Business School Press, Boston Mass.

Christensen, C.M., Suarez, F.F. and J.M. Utterback, 1998. Strategies for survival in fast-changing industries. Management Science 44 (12): S207-S220.

Christensen, C.M., Olesen, M.H. and J.S. Kjaer, 2005. The industrial dynamics of Open Innovation –Evidence form the transformation of consumer electronics. Research Policy 34 (10): 1533-1549.

CIAA, 2005. Data & trends of the European food and drink industry 2004. Confédération des industries agro-alimentaire de l'UE/Confederation of the food and drinks industries of the EU, Brussels.

CIAA, 2006. Data & trends of the European food and drink industry 2005. Confédération des industries agro-alimentaire de l'UE/Confederation of the food and drinks industries of the EU, Brussels.

CIAA, 2007. Data & trends of the European food and drink industry 2006. Confédération des industries agro-alimentaire de l'UE/Confederation of the food and drinks industries of the EU, Brussels.

CIAA, 2008. Data & trends of the European food and drink industry 2007. Confédération des industries agro-alimentaire de l'UE/Confederation of the food and drinks industries of the EU, Brussels.

CIAA, 2009. Data & trends of the European food and drink industry 2008. Confédération des industries agro-alimentaire de l'UE/Confederation of the food and drinks industries of the EU, Brussels.

Clark, P., 1989. Innovation in Technology and Organization. Routledge, London.

Cohen, W., 1995. Empirical studies of innovative activity. In: P. Stoneman (ed.), Handbook of the economics of innovation and technological change. Blackwell Publishers, Oxford.

Cohen, W.M. and D.A. Levinthal, 1990. Absorptive capacity: a new perspective on learning and innovation. Administrative Science Quarterly 35: 128-152.

Combs, K.L. and A.N. Link, 2003. Innovation policy in search of an economic foundation: The case of research partnerships in the United States. Technology Analysis and Strategic Management 15 (2): 177-187.

Commission, 2008. European Innovation Scoreboard 2007. Commission of the European Communities, Brussels.

Conant, J.S., Mokwa, M.O. and P. Varadarajan, 1990. Strategic types, distinctive marketing competencies and organisational performance: A multiple measures-based study. Strategic Management Journal 11: 365-383.

Connor, J.M. and W.A. Schiek, 1996. Food Processing. An Industrial Powerhouse in Transition. Wiley, Chichester.

Coombs, R., Narandren, P. and A. Richards, 1996. A literature-based innovation output indicator. Research Policy 25: 403-413.

Cooper, R.G., 1975. Why New Industrial Products Fail. Industrial Marketing Management 4 (6): 315-26.

Cooper, R.G., 1979. Identifying industrial new product successes: project NewProd. Industrial Marketing Management, 8: 124-135

Cooper, R.G., 1980. Project NewProd: factors in new product success. European Journal of Marketing 14 (5/6): 277-291

Cooper, R.C., 1983. New products do succeed. Research Management 26: 20-25.

Cooper, R.C., 1985. Selecting winning new product projects; using the NewProd system. Journal of Product Innovation Management 2: 34-33.

Cooper, R.C., 1993. Winning at New products. Accelerating the process from idea to Launch (2 ed.). Addison-Wesley Publishing Company, Massachusetts.

Cooper, R.C., 1994. Thrid-generation new product success. Journal of Product Innovation Management 11(1): 3-14.

Cooper, R.C. and J.D. Hlafeck, 1975. Why industrial new products fail. Industrial Marketing Management 4: 315-326.

Cooper, R.G. and E.J. Kleinschmidt, 1987. Success factors in product innovation. Industrial Marketing Management 16: 215-223.

Cooper, R.G. and E.J. Kleinschmidt, 1990. New Products, the Key factors in Success. American Marketing Association, Chicago, pp. 1-50.

Costa, A.I.A., 2003. New insights into consumer-oriented food products design. PhD thesis, Wageningen University, the Netherlands.

Costa, A.I.A. and W.M.F. Jongen, 2006. New insights into consumer-led food product development. Trends in Food Science & Technology, 17: 457-465.

Cottam, A., Ensor, J. and C. Band, 2001. A benchmark study of strategic commitment to innovation. European Journal of Innovation Management, 4: 88-94.

Dahan, E. and J.R., Hauser, 2002. Product development - managing a dispersed process. In: B.A. Weitz and R. Wesley (eds.), Handbook of Marketing, Sage, London. pp. 179-222.

De Vaan, M.J.M., Sneep, C.A.G., Drukker, F.A. and S. ten Have, 1998. Strategische dialoog. Best practices in strategievorming; de onderneming aan zet. Delwell Uitgeverij B.V., 's-Gravenhage.

Dittrich, K. and G. Duysters, 2007. Networking as a means to strategy change: The case of open innovation in mobile telephony. Journal of Product Innovation Management 24 (6): 510-521.

Dosi, G., 1988. The nature of innovative processes. In: Dosi, G., Freeman, C., Nelson, R., Silverberg, G. and L. Soete, 1988. Technical Change and Economic Theory. Frances Pinter, London, pp. 221-238.

Dosi, G., Pavitt, K. and L. Soete, 1990. The economics of technological change and international trade. Harvester Wheatsheaf, UK.

Earle, M.D., 1997. Innovation in the food industry. Trends in Food Science and Technology 8: 166-175.

Earle, M., Earle, R. and A. Anderson, 2001. Food product development. Woodhead Publishing Limited, Cambridge, UK.

ECR Europe, 1999. Efficient product introductions: the development of value-creating relationships. Report of the findings of the Efficient Consumer Response Europe initiative. Ernst & Young and AC Nielsen.

Edquist, C. (ed.), (1997) Systems of Innovation: Technologies, Institutions and Organisations. Pinter Publishers, London.

Enzing, C.M. Lekkerkerk, A., Van Dalen, J.H. and B. Hilberts, 1996. Innovatie in MKB Agribusiness. Deelproject 1: 'Bronnen van Innovatie en knelpunten, TNO-STB/NEHEM Consulting Group, Apeldoorn/'s Hertogenbosch.

Enzing, C.M. and A. van der Giessen, 2005. Technical aspects of the production of functional food ingredients, TNO-STB. Contribution to: Stein, A.J. and E. Rodríguez-Cerezo, (eds.), 2008. Functional food in the European Union, European Commission JRC IPTS, Seville.

Enzing, C.M., Gijsbers, G. and W. Vullings, 2005. Looking forward enriches the future. An essay on influential trends in the agrofood sector and rural areas (in Dutch), Innovatienetwerk Agrocluster en Groene Ruimte, Utrecht,

EIM, 2008. Ondernemen in de sectoren. Feiten en Ontwikkelingen 2007-2009, EIM, Zoetermeer.

Eisenhardt, K. and C. Schoonhoven, 1996. Resource-based view of strategic alliance formation: strategic and social effects in entrepreneurial firms. Organization Science 7: 136-150.

Fagerberg, J., 1995. User – Producer Interaction, Learning and Comparative Advantage. Cambridge Journal of Economics 19: 243–256.

Ferrier, G.D. and P.K. Porter, 1991. The productive efficiency of US milk processing cooperatives. Journal of Agricultural Economics 42: 1-19.

Fetterhoff, T.J. and D. Voelkel, 2006. Managing open innovation in biotechnology. Research - Technology Management 49 (3): 14-18.

Freeman, C. 1982. The economics of industrial innovation (2 ed.). Frances Pinter, London.

Freeman, C. Robertson, A.B., Achilladelis, B.G. and P. Jervis, 1972. Success and failure in industrial innovation. Report on Project SAPPHO by the Science Policy Research Unit. Centre for the Study of Industrial Innovation, University of Sussex, Brighton.

Freeman, C. and L. Soete, (eds.), 1997. The Economics of Industrial Innovation (3 ed.). Pinter, London.

Fuller, G.W., 2005. New food product development: from concept to marketplace (2 ed.). CRC Press LLC, Boca Raton, Florida.

Galazzi, G. and L. Venturini, 1996. Product Innovation in the Food Industry: Nature, Characteristics and Determinants. In: Galazzi, G. and L. Venturini, (eds.), Economics of innovation: the case of the food industry. Physica-Verlag, Heidelberg, pp.133-156.

Galende, J. and J.M. de la Fuente, 2003. Internal factors determining a firm's innovative behaviour. Research Policy 32: 715-736.

Gallo, A.E., 1995. Are there too many new product introductions in the US food marketing? Journal of Food Distribution Research 1: 9-13.

Gilpin, J. and W.B. Traill, 1999. Manufacturer strategies in the European Food Industry. In: G. Galizzi and L. Venturini (eds.), Vertical relationships and coordination in the food system, Physica-Verlag, Heidelberg, pp. 285-310

Gomes-Casseres, B., 1996. The alliance revolution: the new shape of business rivalry. Harvard University Press, Cambridge, MA.

Göransson, G. and E. Kuiper, 1997. Skånemejerier: functional foods through research. In: Traill, B. and K.G. Grunert, (eds.), Product and process Innovation in the food industry. Blackie Academic & Professional, London, pp. 163-174.

Griffin, A. and A.L. Page, 1996. PDMA Success Measurement Project: Recommended measures for product development success and failure. Journal of Product Innovation Management, 13: 478-496.

Grigg, D., 1995. The geography of food consumption: A review. Progress in Human Geography 19: 338-354.

Grunert, K., Larsen, H.H., Madsen, T.K. and A. Baadsgaard, 1996. Market Orientation in Food and Agriculture. Kluwer Academic Publishers, New York.

Grunert, K.G., Harmsen, H., Meulenberg, M., Kuiper, E., Ottowitz, T., Declerck, F., Traill, B. and G. Goransson, 1997. A framework for analysing innovation in the food sector. In: Traill, B. and K.G. Grunert, (eds.), Product and process innovation in the food industry. Blackie Academic & Professional, London. pp. 1-37.

Grunert, K.G., Boutrup-Jensen, B., Sonne, A-M., Brunsø, K., Byrne, D.V., Clausen, C., Friis, A., Holm, L., Hyldig, G., Kristensen, N.H., Lettl, C. and J. Scholdere, 2008. User-oriented innovation in the food sector: relevant streams of research and an agenda for future work. Trends in Foods Science & Technology 19: 590-602.

Hagedoorn, J., 1993. Understanding the rationale of strategic technology partnering: inter-organizational modes of cooperation and sectoral differences. Strategic Management Journal 14: 371-385.

Hagedoorn, J., 2002. Inter-firm R&D partnerships: an overview of major trends and patterns since 1960. Research Policy 31: 477-492.

Hair, J.F., Anderson, R.E., Tatham, R.L., and W.C. Black, 1998. Multivariate data analysis (5 ed.). Prentice-Hall, Inc., New Jersey.

Hakansson, H., 1995. Product development in networks. In: Ford, D. (ed.), Understanding Business Markets: Interaction, relationship and networks. The Dryden Press, New York, pp. 487-507.

Hamel, G., 1991. Competition for competence and inter-partner learning within international strategic alliances. Strategic Management Journal 12: 83-103.

Hamel, G., Y. Doz and C. Prahalad, 1989. Collaborate with your competitors and win. Harvard Business Review Jan-Febr: 133-139.

Hamel, G. and C.K. Prahalad, 1994. Competing for the future. Harvard Business School Press, Boston, Mass.

Hambrick, D., 1983. Some tests of the effectiveness and functional attributes of Miles and Snow's strategic types. Academy of Management Journal 26: 5-26.

Harrigan, K.R., 1985. Strategies for Joint Ventures. Lexington Books, Lexington, Mass.

Hatzichronoglou, T., 1997. Revision of the High-Technology Sector and Product Classification. OECD Science, Technology and Industry Working Papers. 1997/2, OECD Publishing, Paris.

Henderson, R. and K. Clark, 1990. Architectural innovation: the reconfiguration of existing product technologies and the failure of established firms. Administrative Science Quarterly 35: 317-341.

Henderson, R. and I. Cockburn, 1996. Scale, scope and spillovers: the determinants of research productivity in drug discovery. Rand Journal of Economics 27: 32-59.

Herrman, R., 1997. The distribution of product innovations in the food industry: economic determinants and empirical tests for Germany. Agribusiness. 13, 319-334.

Hill, T., 1993. Manufacturing strategy (2 ed.). Macmillan, London.

Hirsch-Kreinsen, H., Jacobson, D., Laestadius, S. and K. Smith, 2005. Low and medium technology industries in the knowledge economy: the analytical issues. In: Hirsch-Kreinsen, H., Jacobson, D., Laestadius, S. (eds.) Low-tech innovation in the knowledge economy. Peter Lang, Frankfurt am Main, pp. 11-30.

Hirsch-Kreinsen, H., Jacobson, D. and P.L. Robertson, 2006. 'Low-tech' industries: innovativeness and development perspectives – a summary of a European research project. Prometheus 24: 3-21.

Hoban, T.J., 1998. Improving the success of new product development. Food Technology 52: 46-49.

Hollanders, H. and A. Arundel, 2005. European Sector Innovation Scoreboards. Technical Paper. European Commission, Brussels.

Hopkins, D.S., 1981. New-product winners and losers. Research management 24 (3): 12-17.

Horowitz, I., 1962. Firm size and research activity. Southern Economic Journal 29: 298-301.

Howell, J.M. and C. Higgins, 1990. Champions of technological innovation. Administrative Science Quarterly 35: 317-341.

Hultink, E.J. and H.S.J. Robben, 1995. Measuring new product success: the difference that time perspective makes. Journal of Product Innovation Management 12: 392-405.

Huston, L. and N. Sakkab, 2006. Connect and develop: inside Procter and Gambler's new model of innovation. Harvard Business Review 84: 58-66.

Hyvönen, S., 1993. The development of competitive advantage. The Helsinki School of Economic and Business Administration.

Imai, K. and Y. Baba, 1989. Systemic Innovation and Cross-Border Networks: Transcending Markets and Hierarchies to create a new techno-economic system. OECD Conference on Science, Technology and Economic Growth. Paris.

INNOVA, 2008. Sectoral Innovation Systems in Europe: The Case of Food, Beverage and Tobacco Sector. Sector report of the Europe Innova project, prepared for the European Commission, Brussels.

Jacobs, D., 1999. Het Kennisoffensief. Slim concurreren in de Kenniseconomie (2 ed.). Samson, Deventer/Alphen aan de Rijn.

Jansen, J.J.P., Van den Bosch, F.A.J. and H.W. Volberda, 2006. Exploratory innovation, exploitative innovation and performance effects of organisational antecedents and environmental moderators. Management Science 52 (1): 1661-1674.

Joppen, L., 2004. New product Development Survey; Faster, higher and stronger. Food Engineering & Ingredients. September: 30-35.

Kanters, R., 1984. The change masters. Irwin, London.

Katz, F., 1998. How major core competencies affect development of hot new products. Food Technology 52: 48-52.

Kearns, G.S., 2005. An electronic commerce strategic typology: insights form case studies. Information & Management 42 (7): 1023-1036.

Ketchen, Jr., D.J., Combs, J.G., Russell, C.G., Shook, C., Dean, M.A., Runge, J.,Lohrky, F.T., Naumann, S.E., Haptonstahl, D.E., Baker, R., Beckstein, B.A., Handler, C., Honig, H. and S. Lamoureux, 1997. Organisational configurations and performance: a meta-analysis. Academy of Management Journal 40: 1-18.

Kirner, E., Kinkel, S. and A. Jaeger, 2009. Innovation paths and the innovation performance of low-technology firms – An empirical analysis of German industry. Research Policy 38: 447-458.

Kleinknecht, A., Poot, T.P. and J.O.N. Reijnen, 1991. Formal and informal R&D and firm size: Survey results from the Netherlands. In: Acs, Z.I. and D.B. Audretsch, Innovation and technological change: an international comparison. Harvester Wheatsheaf, New York.

Kleinknecht, A., van Montfort, K. and E. Brouwer, 2002. The non-trival choice between innovation indicators. Economics of Innovation and New Technology 11(2): 109-121.

Kleinschmidt E.J. and R.G. Cooper, 1991. The impact of product innovativeness on performance. Journal of Product Innovation Management 8: 240-251.

Klevorick, A.K., Levin, R.C., Nelson, R.R. and S.G. Winter, 1995. On the source and significance of interindustry differences in technological opportunities. Research Policy 24: 185–205.

Kline, S.J. and N. Rosenberg, 1986. An Overview of Innovation. In: Landau, R. and N. Rosenberg, (eds.), The Positive Sum Strategy. National Academy Press, Washington DC.

Knox, B., Parr, H. and B. Bunting, 2001. Model of 'best practice' for the food industry. Proceedings of the British Nutrition Society 60, 169a.

References

Knudsen, M.P., 2007. The relative importance of interfirm relationships and knowledge transfer for new product development success. Journal of Product Innovation Management 24 (2): 117-138.

Kogut, B., 1988. Joint ventures: theoretical and empirical perspectives. Strategic Management Journal 9: 319-332.

Kogut, B., Shan, W. and G. Walker, 1992. The make-or-cooperate decision in the context and an industrial network. In: Nohria, N. and R. Eccles, (eds.), Networks and Organizations: structure, forms and actions. HBS Press, Boston MA, pp. 348-365.

Kristensen, K. P. Oostergaard and H.J. Juhl, 1998. Success and failure of product development in the Danish Food sector. Food Quality and Preference 9: 333-342.

Lagnevik, M., Sjöholm, I., Lareke, A. and J. Östberg, 2004. The dynamics of innovation clusters. A study of the food industry. Edward Elgar Publishing, Cheltenham, UK.

LEI, 2005. Landbouweconomisch Bericht 2005. Berkhout, P. and C. Van Bruchem, (eds.), Landbouwkundig Economisch Instituut, Den Haag.

LEI, 2006. Landbouweconomisch Bericht 2006. Berkhout, P. and C. Van Bruchem, (eds.), Landbouwkundig Economisch Instituut, Den Haag.

LEI, 2007. Landbouweconomisch Bericht 2007. Berkhout, P. and C. Van Bruchem, (eds.), Landbouwkundig Economisch Instituut, Den Haag.

LEI, 2008. Landbouweconomisch Bericht 2008. Berkhout, P. and C. Van Bruchem, (eds.), Landbouwkundig Economisch Instituut, Den Haag.

Lester, D.H., 1998. Critical success factors for new product development. Research Technology Management 41 (1): 36-43.

Leonard-Barton, D., 1995. Wellsprings of Knowledge: Building and sustaining the sources of Innovation. Harvard Business School Press, Boston.

Levitt, B. and J. March, 1988. Organisational Learning. Annul Review of Sociology 14: 319-340.

Liebeskind, J.P., Lumerman Oliver, A., Zucker, L. and M. Brewer, 1996. Social networks, learning and flexibility: Sourcing scientific knowledge in new biotechnology firms. Organisation Science 7 (4): 482-443.

Link, P., 1987. Keys to new products success and failure. Journal of Industrial Marketing Management 16: 109-118.

Lord, J.B., 2000. New Products failure and success. In: Brody, A.L. and J.B. Lord, (eds.), Developing new products for a changing market place. Technolmic, Lancaster, pp. 55-96.

Luchs, B., 1990. Quality as strategic weapon. European Business Journal 2: 34-47.

Lundvall, B.A., 1988. Innovation as an interactive process - from user-producer interaction to the National System of Innovation. In: Dosi, G., C. Freeman, R. Nelson, G. Silverberg, and L. Soete (Eds.), Technical Change and Economic Theory. Pinter, London, pp. 349-369.

Lundvall, B.A., 1992. User-Producer Relationships. National Systems of Innovation and Internationalisation. In: Lundvall, B.A., (ed.), National Systems of Innovation, Towards a Theory of Innovation and Interactive Learning. Pinter Publishers. London.

Maidique, M.A. and B.J. Zirger, 1984. A study of success and failure in product innovation. The case of the U.S. electronics industry (Stanford Innovation project). IEEE Transactions on Engineering Management 31: 192-203.

Malerba, F., 2002. Sectoral systems of innovation and production. Research Policy 32: 247–264.

Malerba, F., 2005. Sectoral systems. How and why innovation differs across sectors. In: J. Fagerberg, D. Mowery and R. Nelson (eds.), The Oxford Handbook of Innovation. Oxford University Press, Oxford, pp. 380-406.

Mann, R., Adebanjo, O. and D. Kehoe, 1999. An assessment of management systems and business performance in the UK food and drinks industry. British Food Journal 101: 5-21.

Man, A.P., de, and G. Duysters, 2005. Collaboration and innovation: a review of the effects of mergers, acquisitions and alliances on innovation. Technovation 25 (12): 1377-1387.

Mansfield, E. and S. Wagner, 1975. Organizational and strategic forces associated with probabilities of success in industrial R&D. The Journal of Business 48: 179-198.

Mark-Herbert, C., 2004. Innovation of a new product category - functional foods. Technovation 24; 713-719.

Martinez, M.G. and J. Briz, 2000. Innovation in the Spanish food and drink industry. International Food and Agribusiness Management Review 3: 155-176.

McDaniel, S.W. and J.W. Kolari, 1987. Marketing strategy implications of the Miles and Snow typology. Journal of Marketing 51: 19-30.

McGee, J. and S. Segal-Horn, 1992. Will there be a European food processing industry? In: S. Young and J. Hamill (eds.), Europe and the Multinationals. Edward Elgar, Aldershot.

McKelvey, M., 1996. Evolutionary Innovation: Early Industrial Use of Genetic Engineering. Oxford University Press, Oxford.

Menrad, K., 2004. Innovations in the food industry in Germany. Research Policy 33: 845–878.

Miles, R.E. and C.C. Snow, 1978. Organisational strategy, structure and process. McGraw-Hill, New York.

Miles, R.E., Snow, C.C., Meyer, A.D. and H.J. Coleman Jr, 1978. Organizational strategy, structure, and process. Academy of Management Review 3 (3): 546-562.

Mintzberg, H., 1990. Strategy Formation: Schools of Thought. In: J.W. Frederickson, (ed.), Perspectives on strategic management. Harper & Row, New York.

Mintzberg, H. and J.B. Quinn, 1991. The strategy process: concepts, contexts, cases. Prentice Hall, Englewood Cliffs.

Miotti, L. and F. Sachwald, 2003. Co-operative R&D: why and with whom? An integrated framework for analysis. Research Policy 32 (8): 1481-1499.

Montalvo, C., Tang, P., Mollas-Gallart, J., Vivarelli, M., Marsilli, O., Hoogendorn, J., Leijten, J., Butter, M., Jansen G. and A. Braun, 2007. Driving factors and challenges for EU industry and the role of R&D and innovation. European Techno-Economic Policy Support Network, Brussels.

Montoya-Weiss, M.M. and R. Calantone, 1994. Determinants of new product performance: a review and meta-analysis. Journal on Product Innovation Management 11 (5): 397-417.

Moore, M., 2005. Towards a confirmatory model of retail strategy types: an empirical test of Miles and Snow. Journal of Business Research 58 (5): 696-704.

Mowery D.C, Oxley, J.E. and B.S. Silverman, 1998. Technological overlap and interfirm cooperation: Implications for the resource-based view of the firm. Research Policy 27: 507-523.

Murphy, G.B., Trailer, J.W. and R.C. Hill, 1996. Measuring performance in entrepreneurship research. Journal of Business Research 36: 15-33.

Nahapiet, J. and S. Ghosdal, 1998. Social capital, intellectual capital and organizational advantage. Academy of Management Review 23: 242-266.

Nelson, R.R. (ed.), 1993. National Innovation Systems: A Comparative Analysis. Oxford University Press, New York.

NRC Focus, 2009. Kwartaal uitgaven.

OECD, 2005. Oslo Manual. Guideline for collecting and interpreting innovation data. OECD and Eurostat, Paris

OECD, 2007. Science, Technology and Industry Scoreboard 2007. OECD, Paris.

Oldenziel, R., 2001. Huishouden. In: Schot, J.W., Lintsen, H.W., Rip, A. and A.A. Albert de la Bruhèze, (eds.), Techniek in Nederland in de Twintigste Eeuw, Deel IV: Huishouden, Medische Techniek. Walburg Pers. Stichting Historie der Techniek, Zutphen, pp. 1-152.

Omta, S.W.F., Visscher, E.J., Kemp, R.G.M. and E.F.M. Wubben, 2003. Innovatie in de voedingsmiddelenindustrie – Een verkennende studie naar de kritische succesfactoren voor innovatie in vier agrovoedingsketens. Innovatienetwerk Groene Ruimte en Agrocluster, Utrecht.

Ortt, R.J. and J.P.L. Schoormans, 1993. Consumer research in the development process of a major innovation. Journal of the Marketing Research Society 35: 375-387.

Oudshoorn, N. and T. Pinch, 2003. How users matter: the co-construction of users and technologies. MIT Press, Cambridge.

Palmberg, C., Leppälahti, Lemola, T. and H. Toivanen, 1999. Towards a better understanding of innovation and industrial renewal in Finland – a new perspective. Working Paper 41/99. VTT, Finland

Parr, H, Knox, B. and J. Hamilton, 2001. Problems and pitfalls in the food product development process. The Food Industry Journal 4: 50-60.

Pavitt, K., 1984. Sectoral Patterns of Innovation. Research Policy 13: 343-373.

Peneder, M., 2007. Entrepreneurship and technological innovation. An integrated taxonomy of firms and sectors. Europe Innova Sectoral Innovation Watch, Deliverable WP4. European Commission, Brussels.

Penrose, E., 1959. The theory of the growth of the firm. Basil Blackwell, London.

Peteraf, M., 1993. The cornerstone of competitive advantage: a resource-based view. Strategic Management Journal 14(3): 179-191.

Peters, T.J. and R.W. Waterman, 1982. In Search of Excellence. Warner, New York.

Petroni, A. and B. Panciroli, 2002. Innovation as a determinant of suppliers' roles and performances: an empirical study in the food machinery industry. European Journal of Purchasing & Supply Management 8: 135-149.

Pisano, G. and D. Teece, 1989. Collaborative arrangements and global technology strategy: some evidence from the telecommunications equipment industry. In: R. Rosenbluhm (ed.), Research on Technological Innovation, Management and Policy, 4, JAI Press. Greenwich, CT. pp. 227-256.

Porter, M., 1980. Competitive strategy: techniques for analysing industries and competitors. Free Press, New York.

Porter, M., 1985. Competitive advantage, creating and sustaining superior performance. Free Press, New York.

Porter, P.K. and G.W. Scully, 1987. Economic efficiency in cooperatives. Journal of Law and Economics 30: 489-512.

Rama, R., 1996. Empirical Study on Sources of Innovation in International Food and Beverage Industry. Agribusiness 12 (2): 123-134.

Rama, R., 1998. Growth in Food and Drink Multinationals, 1977-94: An Empirical Investigation. Journal of International Food and Agribusiness Marketing 10 (1): 31-52.

Reijnders, L. and R. Sijmons, 1974. Voedsel in Nederland. Van Gennep, Amsterdam.

Reinstaller, A. and F. Unterlass, 2008. What is the right strategy for more innovation in Europe? Drivers and challenges for innovation performance at the sector level, Europe Innova Sectoral Innovation Watch, Synthesis report. WIFO, Vienna.

Rigby, D. and C. Zook, 2002.Open-market innovation. Harvard Business Review 80 (10): 80-89.

Robertson, P.L., Pol, E. and P. Carroll, 2003. Receptive capacity of established industries as a limiting factor in the economy's rate of innovation. Industry and Innovation 10: 457-474.

Robertson, P.L. and P. Patel, 2007. New wine in old bottles: technological diffusion in developed economies. Research Policy 36: 708-721.

Robertson, P.L., Smith, K. and N. von Tunzelmann, 2009. Innovation in low- and medium-technology industries. Research Policy 38: 441-446.

Rochford, L. and W. Rudelius, 1997. New product development process; stage and success. Marketing Management 26: 67-84.

Rosenberg, N., 1963. Technological change in the machine-tool industry, 1840-1910. Journal of Economic History 23: 414-446.

Rosenberg, N., 1976. Perspectives on Technology. Cambridge University Press, Cambridge.

Rothwell, R., 1976. The 'Hungarian SAPPHO': some comments and comparisons. Research Policy 3: 30-38.

Rothwell, R., 1992. Successful industrial innovation: critical success factors for the 1990s. R&D Management 22: 221-239.

Rothwell, R., 1994. Towards the Fifth-generation Innovation Process. International Marketing Review 11 (1): 7-31

Rothwell, R., Freeman, C., Horsley, A., Jervis, V.T.P., Robertson, A.B. and J. Townsend, 1974. SAPPHO Updated - project SAPPHO Phase II. Research Policy 3:259-291.

Rource, J.B. and R.H. Keeley, 1990. Predictors of success in new technology based ventures. Journal of Business Venturing 5: 221-239.

Rozin, P., 1987. Psychological perspectives on food preferences and avoidances. In: Harris, M. and E.B. Ross, (eds.), Food and Evolution. Towards a theory of human food habits. Temples University Press, Philadelphia.

Rubin, P.H., 1973. The expansion of firms. Journal of political Economy 81: 936-949.

Rudder, A., Ainsworth, P. and D. Holgate, 2001. New Food product development: strategies for success. British Food Journal 103: 657-671.

Rudolph, M.J., 1995. The food product development process. British Food Journal 97: 3-37.

Rumelt, R.P., 1974. Strategy, structure and economic performance. Harvard University, Boston M.A.

Rumelt, R.P., 1982. Diversification strategy and profitability. Strategic Management Journal 3: 359-369.

References

Saguy, I.S. and H.R. Moscowitz, 1999. Integrating the consumer into new product development. Food technology 53: 68-73.

Sanders, G. and B. Neuijen, 1987. Bedrijfscultuur: Diagnose en Beïnvloeding, Van Gorcum, Assen.

Sandven, T., Smith, K. and A. Kaloudis, 2005. Structural change, growth and innovation: the role of medium and low-tech industries, 1980-2000. In: Hirsch-Kreinsen, H., Jacobson, D. and S. Laestadius, (eds.), Low-tech Innovation in the Knowledge Economy. Peter Lang, Frankfurt-am-Main, pp. 31-59.

Santamaría, L., Nieto, M.J. and A. Barge-Gil, 2009. Beyond formal R&D: taking advantages of other sources of innovation in low- end medium-technology industries. Research Policy 38: 507-517.

Sarkar, S., and A.I.A. Costa, 2008. Dynamic of open innovation in the food industry. Trends in Food Science & Technology 19: 574-580.

Scherer, F.M., 1982. Inter-industry technology flows in the United States. Research Policy 11: 227-245.

Scherer, F.M., 1984. Corporate Size, diversification and innovative activity. In: F. Scherer, (ed.), Innovation and Growth. A Schumpeterian Perspective. MIT Press, Cambridge, MA, pp. 222-238.

Schumpeter, J.A., 1939. Business Cycles: A theoretical, historical and statistical analysis of the capitalist process. McGraw Hill, New York & London.

Schmidt, J.B., 1995. New product myopia. Journal of Business & Industrial Marketing 10 (1): 23-33.

Schmierl, K. and H. Köhler, 2005. Organisational learning: knowledge management and training in low-tech and medium low-tech companies. ournal for Perspective on Economic Political and Social Integration 11 (1-2): 171-221.

Selnik, 1957. Leadership in Administration. Harper & Row, New York.

Sivadas, E. and F.R., 2000. An examination of organisational factors influencing new product success in internal and alliance-based processes. Journal of Marketing 64: 1-49.

Smallbone, D., North, D. and R. Leigh, 1993. The use of external assistance by mature SMEs in the UK: some policy implications. Entrepreneurship & Regional Development 5: 279-295.

Smit, G., Fortuin, F.T.J.M. and S.W.F. Omta, 2008. Open innovation in the food industry; the Unilever flavour experience. Paper for the 8th International conference on management in agrifood chains and networks. Ede.

Snow, C.C. and L.G. Hrebinaik, 1980. Strategy, Distinctive competence and organisational performance. Administrative Science Quarterly 25: 317-335.

Souder, W. and J. Sherman, 1994. Managing new technology development. McGraw-Hill, New York.

Steenkamp, J.B.E.M. and H.C.M. Van Trijp, 1996. Quality guidance: a consumer-based approach to food quality improvement using partial least squares. European Review of Agricultural Economics 23 (2): 195-215.

Stein, A.J. and E. Rodríguez-Cerezo, (eds.), 2008. Functional food in the European Union, a report based on the ESTO study Functional Food in European Union, European Commission JRC IPTS, Seville.

Stewart-Knox, B. and P. Mitchell, 2003. What separates the winners from the losers in new food product development? Trends in Food Science & Technology 14: 58-64.

Stewart-Knox, B., Parr, H., Bunting, B. and P. Mitchell, 2003. A model for reduced fat food product development success. Food Quality and Preference 14: 583-593

Stewart, H. and S. Martinez, 2002. Innovation by food companies key to growth and profitability. Food Review 25 (1): 28-32.

Teece, D.J., 1982. Towards an economic theory of the multiproduct firm. Journal of Economic Behaviour and Organisation 3: 39-63.

Teece, D.J., 1986. Profiting from technological innovation: implications for integration, collaboration, licensing and public policy. Research Policy 15: 285-305.

Teece, D.J. and G. Pisano, 1994. The dynamic capabilities of firms: An introduction. Industrial and Corporate Change 3 (3): 537-556.

Teece, D.J., Pisano, G. and A. Shuen, 1997. Dynamic Capabilities and Strategic Management. Strategic Management Journal 18: 509 –533.

Tidd, J., Bessant, J. and K. Pavitt, 2005. Managing Innovation. Integrating Technological, Market and Organizational Change (3 ed.). John Wiley & Sons, Chichester.

Traill, B., 2000. Strategic groups in the F&B industry, Journal of Agricultural Economics 52 (1): 45-60.

Traill, B. and K.G. Grunert, (eds.), 1997. Product and process innovation in the food industry. Blackie Academic & Professional, Chapmann & Hall, London.

Traill, W.B. and M. Meulenberg, 2002. Innovation in the food industry. Agribusiness 18: 1–21.

Urban, G.L. and J.R. Hauser, 1993. Design and marketing of new products (2 ed.). Prentice-Hall, Englewood Cliffs NJ.

Uzzi, B., 1997. Social structure and competition in interfirm networks: the paradox of embeddedness. Administrative Science Quarterly 42: 35-67.

Van der Panne, G., 2004. Entrepreneurship and localized knowledge spillovers. PhD Thesis. Technical University Delft, Delft.

Van der Panne, G., 2007. Issues in measuring innovation. Scientometrics 71: 495-507.

Vanhaverbeke, W. and M. Cloodt, 2006. Open innovation in value networks. In: Chesbrough, H.W., Vanhaverbeke, W. and J. West, (eds.), Open innovation: Researching a new paradigm. Oxford University Press, Oxford, pp. 258-281.

Van Otterloo, A.H., (ed.), 2000. Voeding. In: Techniek in Nederland in de twintigste eeuw. Deel III. Landbouw en Voeding. Stichting Historie der Techniek. Walburg Pers, Zutphen, pp. 235-374.

Van Poppel, A., 1999. Nieuwe producten. Te veel missers. Trends May, 13th, 78-79.

Van Reenen, G.J. and B. Waisfisz, 1995. Organisatiecultuur als beleidsinstrument. Meten is weten, Stichting Maatschappij en Onderneming, Den Haag.

Van Trijp, H.C.M. and M.T.G. Meulenberg, 1996. Marketing and consumer behaviour with respect to foods. In: Meiselman, H.L. and H.J.H. MacFie, (eds.), Food Choice, Acceptance and Consumption. Blackie Academic & Professional, London.

Van Trijp, H.C.M. and J.E.M. Steenkamp, 1998. Consumer-oriented new product development: principles and practice. In: Jongen, W.M.F. and M.T.G. Meulenberg, (eds.), Innovation of Food Production Systems: Product Quality and Consumer Acceptance. Wageningen Pers, Wageningen. pp. 37-66.

Van Trijp, H.C.M. and E. Van Keelf, 2008. Newness, value and new product performance. Trends in Food Science & Technology 19 (11): 562-573.

Venkatraman, N. and V. Ramanujam, 1986. Measurement of business performance in strategic research; a comparison of approaches. Academy of Management Review 11: 801-814.

Volberda, H.W., Van den Bosch, F.A.J. and J.J. Jansen, 2006. Slim managen & Innovatief organiseren. Onderzoeksverslag. Eiffel. Rotterdam.

Von Hippel, E., 1988. The Sources of Innovation. Oxford University Press. Oxford.

Warner, M., 1996. Innovation and training. In: Dodgson, M. and R. Rothwell., (eds.), The Handbook of Industrial Innovation. Edward Elgar, Cheltenham, pp. 348-354.

Weber, M., 1947. The theory of social and economic organisation. Oxford University Press, New York.

Weick, K.E. (1979) The social psychology of organisation (2 ed.). Addison-Wesley, Reading, Mass.

Wernerfelt, B., 1984, A resource base view of the firm. Strategic Management Journal 5: 171-180.

Wheelwright, S. and K. Clark, 1992. Revolutionising product development. Free Press, New York.

Wilkinson, J., 1998. The R&D priorities of leading food firms and long-term innovation in the agrofood system. International Journal of Technology Management 16: 711-720.

Wijnands, J.H.M., Van der Meulen, B.M.J. and K.J. Poppe, (eds.), 2006. Competitiveness of the European Food Industry. An economic and legal assessment. Reference no. ENTR/05/75. European Commission, Brussels.

Wind, J. and V. Mahajan, V., 1998. Issues and opportunities in new product development: an introduction to a special issue. Journal of Marketing Research 34: 1-12.

Zirger, B.J., 1997. The influence of development experience and product innovativeness on product outcome. Technology Analysis and Strategic Management 9: 287-297.

Appendix – Overview of the variables and the related questions in the survey

Market performance of the product

The product's short-term market performance was measured using two questions:
Q: What was the impact of the product on the company's market share? (ordinal variable: large increase = 4, small increase = 3, no change = 2, decrease = 1)
Q: What was the impact of the product on the company's turnover? (ordinal variable: large increase = 4, small increase = 3, no change = 2, decrease = 1)
The values for the variable 'short-term market performance' are the average of the scores on impact on market share (Q1) and impact on turnover (Q2).

Long-term market performance was measured using one question:
Q: Is the product still on the market? (dichotomous variable: Yes = 1; No = 0)

Innovativeness of the product

Newness of product was measured using one question:
Q: Is the product a new product or an improved/renewed version of an already existing product? (dichotomous: new product = 1; improved/renewed product = 0;).

Newness of product attributes was measured using eight questions, each dealing with one specific product attribute:
Q: Is the product new or improved/renewed concerning: 1) raw materials; 2) ingredients; 3) processing; 4) recipe; 5) shelf-life conditions; 6) readiness to use/eat; 7) packaging; 8) nutritional value? (dichotomous variables: Yes = 1; No = 0)
The values for the variable 'New Product Attributes' are the sum of the scores on the 'New product attribute' questions Q1-8: the more new product attributes, the higher its innovativeness (ratio scale).

Newness of knowledge used was measured using one question:
Q: Has the new scientific–technological or other knowledge used for the development of the product been developed by the company itself, was it developed by others, or has existing and earlier applied knowledge been used? (ordinal variable: developed by the company itself = 3, developed by others = 2, existing knowledge = 1)

Newness of the production process was measured using one question:
Q: Was a totally new production process developed; or has the existing process drastically or considerably been adjusted or were there no major changes? (ordinal variable: new process = 4, adjusted-significant = 3, adjusted-small = 2, no major change = 1)

Market of the product

Market of the product was measured using three questions:
Q: Is it a product 1) for the consumer market; 2) for the food service market; 3) for the industrial market? (dichotomous variables: Yes = 1, No = 0)

Resources: source of ideas

The use of sources of ideas for the development of the product was measured using the same question for each of the thirteen sources listed:
Q: Has this been a source of ideas for developing the product? 1) Company's R&D department, 2) company's marketing / sales department, 3) company's board of directors, 4) suppliers of machinery & equipment, 5) suppliers of ingredients & raw materials, 6) universities, 7) research institutes, 8) polytechnics (HBO), 9) professional literature, 10) patent literature, 11) customers/consumers, 12) competitors, 13) fairs and exhibitions.
The value for the variable 'source of innovation - research organisations' is based on the scores on the questions for universities, research institutes and polytechnics.
(dichotomous variables: Yes = 1, No = 0).

Resources: company functions

The involvement of company functions in the product innovation process was measured using the same question for each company function listed:
Q: Was the company function involved in the product innovation process? 1) R&D, 2) purchase, 3) quality, 4) production, 5) marketing, 6) sales, 7) logistics (dichotomous variables: Yes = 1, No = 0).

Resources: preparatory and test activities

The use of preparatory and test activities in the product innovation process was measured using the same question for each activity listed:
Q: Has the activity been performed during the product's innovation process? 1) Technical feasibility, 2) check patents and licences, 3) marketing research, 4) check product by potential clients, 5) competitive analysis, 6) test product qualities, 7) test production process, 8) test marketing activities, 9) test product by consumer (dichotomous variables: Yes = 1, No = 0).
The values for the variable 'Scale of preparatory and test activities' is based on the number of different preparatory and test activities that were carried out (ordinal variable).

Resources: external actors

The involvement of external actors as partner in the product innovation process was measured using the same set of questions for each external actor:
Q: Did your company work together with this partner in the product innovation process and if yes, was this on an incidental or a structural base, and on what activities did you work together? 1) Customers, 2) suppliers, 3) competitors, 4) consultancy companies (including engineers, marketers), 5) universities, 6) research institutes, 7) polytechnics (dichotomous variables: Yes = 1, No = 0).
The variable 'partner - research organisations' is based on the scores on the questions for universities, research institutes and polytechnics (dichotomous variable: Yes = 1, No = 0).
In the analysis we only used the data on involvement or non-involvement for the variables 'Partner – xxxxxx'.

The involvement of external actors as outsourcers was measured using the same question for each external actor:
Q: Did you outsource activities in the innovation process of this product to this company/ organisation? 1) Universities, 2) research institutes, 3) polytechnics, 4) companies providing recipe advice, 5) companies providing marketing-related activities (dichotomous variables: Yes = 1, No = 0).
The variable 'outsource - research organisations' is based on the scores on the questions for universities, research institutes and polytechnics (dichotomous variable: Yes = 1, No = 0).

The involvement of external actors as suppliers of goods and services for the product innovation process was measured using the same question for each external actor:
Q: Did you purchase these goods or services for use in the product's innovation process? 1) Machinery and/or equipment, 2) training, 3) licenses/software (dichotomous variables: Yes = 1, No = 0).

Openness of the innovation network

As proxy for openness of the innovation network two variables were created, using data collected through variables on the involvement of external actors:
The values for the variable 'openness - idea stage' is the sum of the number of different external actors that have provided ideas for innovations (ratio scale). The values for the variable 'openness-development stage' is the sum of the number of different external actors involved in the network as partners, outsourcers or sellers (ratio scale).

Appendix

Organisational routines

The use of four organisational routines was measured using the same question:

Q: Has this procedure/activity been used during the product's innovation process? 1) Formalised product development plan, 2) evaluation procedure, 3) go/no-go decision, 4) cross-functional team (dichotomous variables: Yes = 1, No = 0).

The values for the variable 'number of organisational routines' are based on the number of different routines that were used (ratio scale).

Company culture

The company culture was measured using seven questions each addressing one of the seven dimensions of company culture:

Q: How do you assess your company's culture for this characteristic? 1) Flexibility, 2) openness, 3) cooperation between company departments, 4) supportive management style, 5) human focus, 6) focused on results, 7) creativity (interval scale: high = 3, medium = 2, low = 1).

Product innovation strategy

Product strategy was measured for seven strategy-related goals using a combination of two questions:

Q: Was the strategic goal applicable for the company for developing this product? (dichotomous variables: Yes = 1, No = 0).

Q: If 'Yes', how important was the strategic goal? (ordinal variable: very important = 4; important = 3; somewhat important = 2; not important = 1). 1) extend product range, 2) replace products being phased out, 3) improve product quality, 4) create new markets, 5) address a specific market need 6) keep up with competitors, 7) increase/keep market share.

The values for the seven variables combine the scores for the two questions (ordinal variable: strategic goal was applicable and very important = 5; applicable and important = 4; applicable and somewhat important = 3; applicable, but not important = 2; strategic goal was not applicable = 1).

Strategy-related activities were defined as a number of specific activities that were performed in the preparatory phase of the product innovation process (see 'Preparatory and test activities') and the importance of the use of technical findings for the company (questions same as for strategy-related goals).

Company size

Company size was measured using one question:

Q: Which size-category applies to your company: 1-9, 10-49, 50-99, 100-149, 150-249, 250-499 or more than 500 employees?

A dummy variable 'LF' was created (1 for companies with more than 250 employees; 0 for companies with less than 250 employees).

Summary

For companies in an increasing number of industrial sectors, innovation has become essential for their competitive advantage. Studies also show that investments in the development of new knowledge and technology for stimulating innovation-based growth, clearly lead to better economic performance. Changes in customer demands, new scientific and technological developments, increased global competition and changing legislation create new challenges. Through product and process innovation, companies are able to deliver better, faster and cheaper high-quality products and services. For that reason the management of innovation is a core activity of innovating companies.

However, research interest in the innovation behaviour of companies has focused mostly on high-tech industries, while most innovations are developed in low- and medium-tech industries. These studies also focus mainly on the research and development (R&D) activities as the determinant of innovation, while the R&D activities of companies that have no formal R&D department – which is often the case in low and medium-tech industries – are often underestimated. There is a clear lack of understanding of the specific characteristics of innovation processes in low- and medium-tech industries and the role of R&D, while these industries are of comparatively greater importance for national economies than the high-tech sectors. The present study tries to fill this gap by focusing on product innovation in a specific low- to medium-tech industry: the food and beverage industry (F&B).

In recent years, this industry has been confronted with a number of challenges: increased competition especially from retailers' own label products, stricter government regulations concerning food safety and recently also the economic crisis leading to less spending by the consumer. These developments have stimulated the management of many F&B companies to become more pro-active and take more commercial initiatives and one of the best strategies is to be innovative. Studies confirm that both large companies and small and medium companies in the F&B industry have been successful in this respect. Important strategic and operational questions these companies have to answer deal with the choice of a specific product innovation strategy, the partners to co-innovate with and how to manage the innovation process so that all relevant resources are combined successfully. This book tries to answer these questions by discussing the important topic of product innovation in the F&B industry.

The main research question to be answered in this book is:

> *What key factors are positively related to the short- and long-term market performance of F&B products?*

The theoretical framework of this research project is based on concepts in the field of innovation management and concepts that belong to the 'resource-based view'. The research project focuses on the innovative product and is distinct from earlier studies in the field of innovation management that focus mainly on the innovative company.

For the analysis a product-based dataset has been constructed. By screening professional and trade journals for the Dutch F&B industry that were published in the second half of 1998, 200 products were identified, with the name of the company that had developed and marketed the product. In order to assure that the products had survived the critical first year on the market, data were collected in the period March-September 2000. This included data on the innovation process, strategy, internal and external resourcing, product innovativeness and market performance; they were collected through phone interviews based on a structured questionnaire with the managers that had been involved in developing and marketing the products. In the end, complete datasets could be collected for 129 products. The sample of 129 products included 76 products for the consumer market, 37 for the industrial market and 16 for the food service market. An important new asset of our research project is that we also have included a variable for long-term market performance. In December 2005 data collection was completed with the data on the long-term market performance of the products.

In order to answer the main research question four studies have been carried out using this data on F&B products. This first two studies focus on the innovation process. The first study (Chapter 2) aims to understand how the differences in the use of in-house resources in the innovation processes relate to the short- and long-term market performance of new versus improved products. The second study (Chapter 3) explores the differences between the innovation process of products for the consumer and the industrial market. The third study (Chapter 4) investigates the relationship between innovation strategy and product market performance. The fourth study (Chapter 5) focuses on the openness and composition of the innovation network of new and improved products and the short- and long-term market performance of these products.

There is no consensus in literature on the importance of market- versus technology-related resources in the innovation process of new versus improved products. The few studies that address the management of new versus improved product development processes come to different conclusions. Earle *et al.* (2001) found that in the case of new products, product development involves relatively more market-related activities (i.e. activities such as market research, consumer tests etc.) because the market is still unknown. In the case of improved products, the market is known and only adaptations are needed. On the other hand, Balanchandra and Friar (1997) argue that for new products the market still has to be conquered; therefore technology-related activities (i.e. dealing with the technical aspects of the product and its production process) are more important. For the improved product, once it has reached a certain market share, market factors are very important in order to maintain and increase that market share (*ibid.*).

Technology-related resources can be actors (such as research organisations, suppliers of ingredients or technical equipment) or knowledge-carriers (such as scientific publications, patents) from inside and outside the company that provide input of a technological character; this can be knowledge *pur sang* or embedded in products, processes or services. Similar, market-related resources, such as market knowledge, brand names, customer knowledge or downstream actors in the distribution channel contribute market-related input to the innovation process.

The explorative study presented in Chapter 2 aim is to investigate for the groups of new and improved products separately the differences in their product innovation process and how the technology-related and market-related in-house resources used in this process relate to their short and long-term market performance. The study research question is therefore:

> *Which technology- and market-related resources play a key role in the short- and long-term market performance of new versus improved products?*

First of all we found that 64% of the products in our sample were still on the market seven years after product launch. The products in our sample were on the market one and a half years after market launch, which means that they passed the critical first year, where about one out of three newly launched F&B products for the retail market fail. This is one explanation for the unexpectedly high success rate. Also the method we have used for the identification of the products might provide some explanation. We used products that were selected by the editorial board of the professional journals. It can be assumed that this board chose products that they assumed would have a good chance of being successful in the market.

We also found that new products perform better in the short term and improved products in the long term. After having made a distinction between high, medium and low short time performance, we found that new products with a high short-term market performance also perform better in the long term than improved products.

We found that for new product's successful short-term market performance a number of both technology- and market-related in-house resources (activities, company functions involved) are significantly important. The technology-related resources included technical feasibility, test of the production process, and the involvement of the company's production function. The market-related resources included marketing research, test of the product concept and of the product by the customer and the involvement of the company's marketing function For improved products, only one market-related actor is of significant importance: the involvement of the company's sales function. Technology-related resources are much more critical for the improved products' short-term market performance. They include the company's R&D function both as source of ideas and its involvement in the product development process and carrying out technical feasibility studies and the testing of the production process. With respect to long-term market performance it was found that a combination of technology-related resources (R&D function involved and test of product quality) and market-related resources (marketing research and test of product concept with customers) are only crucial for improved products; not for new products. We conclude that – also due to the many large uncertainties that accompany the development of new products – other innovation trajectories and management systems can play a key role in the long-term market performance of these products.

After having explored the specificities of the product innovation process of new versus improved products in Chapter 2, the study in Chapter 3 focuses on the characteristics of the innovation process of consumer versus industrial products. Earlier studies on product innovation in this industry focused only on products for the consumer market or covered all products but did not distinguish between consumer and industrial products. We expect differences in their innovation processes; consequently innovation management may have to focus on different aspects. The descriptive study investigates the role of technology-related and market-related internal and external resources in the innovation process of the two product groups. The research question to be answered is:

> *What are the differences in the involvement of technology-related and market-related resources in the innovation process of consumer versus industrial products and in their short- and long-term market performance?*

We found for consumer products that in the development stage of the product innovation process two market-related resources (more specifically: the company's marketing function and outsourced marketing-related activities) were involved significantly more often compared to industrial products. In the idea stage market-related sources of innovation were used more often in the case of industrial products (this applied to customers and competitors). For technology-related resources we found that in the idea stage they were used more often in the innovation process of industrial products (patent literature). However, in the development stage technology-related resources (suppliers of ingredients and or equipment) were used significantly more often in the case of consumer products. We conclude that both market- and technology-related resources are significantly important in the innovation process of both consumer and industrial products.

Industrial products showed an overall better performance in the long term: 76% of the industrial products were still on the market seven years after product introduction, versus 57% of the consumer products. Industrial products performed better in the long term in all three short-term performance groups (high, medium, low). In the short term consumer products performed better in the high performance group; while industrial products performed better in the medium and low performance groups. We explained the better performance of industrial products, i.e. the ingredients, by the use of ingredients in a multitude of consumer products some of which may have a longer life cycle.

Chapter 4 presents a study on innovation strategies. An articulated innovation strategy is an important factor for the innovative product's success on the market. Such a strategy provides guidelines for questions such as which market to enter, with which products and which resources to develop. As companies in the F&B industry increasingly have to compete on the basis of new and more advanced products, the development of an innovation strategy is becoming increasingly important. We therefore asked the following research question:

> *What is the impact of the product innovation strategy on the product's short- and long-term market performance?*

Miles and Snow (1978) have developed a theoretical framework of strategic archetypes which distinguishes between three types: prospectors, analyser and defenders. The importance of product innovation differs per strategy type and consequently so does the need for (new) resources. Companies that follow a prospector strategy foster growth by developing new products and exploiting new market opportunities. Those following a defender strategy grow mainly through market penetration and by producing and distributing their goods as efficiently as possible. Analyser companies combine the strengths of both the prospector and the defender strategies; they attempt to minimise risks by moving towards new product and markets only after their viability has been demonstrated, while maximising the opportunity for profit by its stable product and market areas. Each type shows a consistent pattern of response to the changing environments in which they operate.

Our findings showed that companies that focused on a prospector strategy in combination with performing extensive market assessment activities in preparing the product development process, are successful in the short term, but not in the long term. Pursuing a defender strategy leads to long-term market performance. Our explanation is that companies that follow a prospector strategy with a high level of innovative activity, have introduced in the period until 2005 new products which have succeeded the products we found in 1998. On the contrary, companies following a defender strategy that have introduced a new product are less likely to replace this; instead, they will try to keep their products on the market for as long as possible. By making regular small product improvements, companies following such a defender strategy aim to maintain and protect the market position of their once introduced new product at a certain level of successful market performance. The results suggest that a more segmented categorisation of company strategies is more appropriate for understanding what strategic factors influence the products' successful market performance.

In Chapter 5 the impact of the innovation network on the product's short- and long-term market performance was investigated. In this chapter we addressed the following research question:

> *What is the impact of the openness and composition of the product-related innovation network on the short- and long-term market performance of the product?*

Summary

Openness refers to 'open innovation' (Chesbrough, 2003) which stands for the increase in the use of external resources by innovative companies in order to speed up the innovation process. Until recently, open innovation studies focused mainly on large and R&D-intensive companies; but this is changing and there is a growing interest in open innovation in low- and medium-tech sectors and in small and medium-sized companies.

Studies already had found that F&B companies rely more on external sources of innovation than the average for all industries. The interactions between companies and their business partners in the supply chain as well as with public research organisations play a crucial role in achieving successful innovations. We found that the more of these external resources were involved in the product-related innovation network (i.e. the more open the network), the better the product's short- and long-term market performance. When investigating the role of product innovativeness, we found that the relationship between openness of the product-related innovation network and the product's market performance was significant for both new and improved products, but most significant for new products.

We expected to find different levels of involvement of external technology- and market-related actors in the product's network and also differences in the impact of these resources on the product's short- and long-term market performance, depending on the innovativeness of the product. For new products we found that both technology- and market-related actors contributed significantly positive to the product's market performance. These actors can be involved as source of innovation in the idea stage and as partner, outsourcer of seller in the development stage of the product innovation process. The customer and competitor as market-related source of ideas, the research organisations to which activities are outsourced and the companies that supply machinery and/or equipment (both technology-related) contributed most significantly to long-term market performance. Most important for the short-term market performance of new products were the companies that supplied ingredients and/or raw materials and machinery and/or equipment as sources of ideas for the new products, the contribution of companies to which market-related activities are outsourced and the companies that provided training and that sold machinery and/or equipment.

For the group of improved products we found that the involvement of technology-related actors, and more specific companies supplying ingredients and/or raw materials as source of innovation and companies that supplied machinery and/or equipment and those that supplied raw materials and/or ingredients as partner of the innovating company strongly negatively affected the long-term market performance of these products. A most probably explanation could be that the relationship with many ingredient and raw material suppliers was stopped – due to the rationalisation of suppliers in the beginning of the 2000s – and the products with these ingredients had been withdrawn from the market. Appropriability problems between the innovating company and its suppliers and a lack of cross-disciplinarity and thus missing conditions for product ideas when linking with partners from one's own industry, could also explain this outcome.

Product innovation in the Dutch food and beverage industry

On the basis of the findings of these four studies we come to a number of conclusions that provide an answer to the central research question of this book.

The results of these studies have revealed important key insights into the role of technology- and market-related resources in successful product innovation in a low- to medium-tech industry, the F&B industry. Our research project is one of the first to investigate the technology-related resources of product innovation in the F&B industry; this subject was more or less neglected. Our study has filled this gap. Although the F&B industry is characterised as a medium-low-tech industry according to formal statistics based on in-house R&D activities, and the focus in literature on innovation in the F&B industry is mainly on the market-driven and user-oriented character of innovation, our findings show that the use of technology-related resources in product innovation was very important for the successful performance of the F&B products on the market, both in the short and long term.

Our study is one of the first on open innovation in a low- to medium-tech industry and on the importance of open innovation for the market performance of products. We found that the involvement of customers in the innovation process does not significantly relate to the products' short- or long-term market performance. This is not what we had expected on the basis of marketing literature on user-oriented product innovation. We found that the customers of ingredient producers (these customers are the producers of consumer products) were significantly more often involved in the innovation process than the customers of the producers of consumer products (retailers). As ingredient producers increasingly develop total product concepts, including recipes and prescriptions for production of the consumer products, we concluded that the close collaboration between the producers of ingredients and that of consumer products could represent another channel through which market knowledge is being transferred.

We also found that the role of suppliers in the F&B industry has changed. Traditionally, the F&B industry mainly innovated through process improvements, facilitated by suppliers of advanced machinery and equipment. The F&B industry, like many other low- and medium-tech industries relied heavily on embodied technologies for improved productivity. However, other studies have shown that product innovation is becoming more important in the F&B industry and our studies show that ingredient suppliers more than suppliers of equipment have been involved in the innovation process. We recommend that the role of suppliers in innovation in the F&B industry be revisited; new types of suppliers (providing ingredients, instead of equipment) are now contributing significantly to product innovation in this industry and their role in the innovation process is changing, illustrating the opening up of the innovation processes.

Our studies have provided important contributions to the literature in the field on innovation management.

Summary

Firstly, by focusing our studies on the innovative product rather than on the company as is the case in most studies, and by distinguishing between new and improved products, we found important new insights into what are the key determinants of successful innovation management of new and improved products.

Most studies on successful innovation management only include the short-term performance of products. An important contribution of our studies is that – by also measuring the product's market performance in the long term – we have gained insight into key factors for innovation management aiming at successful performance in the long term.

A third contribution of our studies is that they provide important insights into the role of the innovation strategy in successful market performance of new and improved products in the short and long term.

Finally, we provided important new insights into product innovation in low- and medium-tech industries and in the role of technology- and market-related resources. These industries are very important for the national economy; the new insights provided by our studies can be used to improve their performance.

About the author

Christina Margaretha Enzing was born in Weststellingwerf on 28th November 1952. In 1981 she received her MSc degree in Chemistry at the University of Groningen, with a major in structure-function relationship of proteins. During her chemistry course her interest in questions concerning the social aspects of science and the responsibility of scientists was raised. She was one of the first students to take a minor in the new field of the Social Studies of Science; for this she analysed the policy aspects of the recombinant DNA debate in the Dutch parliament (late 1970s). She was also involved in the start-up of the so-called Chemistry Shop at the Chemistry Subfaculty of the University of Groningen. As student-assistant she co-developed several training programs in the field of Science and Society for BSc students in chemistry and biology and participated in the teaching of these programs. She was able to pursue her interest in the history, sociology and philosophy of science as member of the editorial board of a student's magazine. In 1981 she became member of the board of the national association of scientific workers (VWW); in this position she organised several conferences (including: Technology Assessment, Advice of the Board DNA Committee) and wrote articles for the VWW journal 'Wetenschap en Samenleving'.

After graduating, Christien worked for another year at the University of Groningen on a study on multidisciplinary research. In 1982 she joined the 'Chemistry and Society' group at the University of Utrecht, where she was involved in a study that explored the possibilities of demand-driven agenda setting for agricultural research. She also taught evening classes in 'Science and Society' for part-time chemistry students.

In 1985 Christien joined the TNO institute for Strategy, Technology and Policy (STB). Initially, she worked for Walter Zegveld in his position as advisor to the Dutch government on technology and innovation policy issues. After this period, her work increasingly focused on biotechnology and life sciences and its applications in specific fields: agriculture, food and non-food (biobased economy). More recently she extended her field of interest to nanotechnology. She has been leading research projects that were performed by (international) multi-disciplinary teams; these included studies on the efficiency of life sciences and biotechnology policies, the economic impact and technology assessment of emerging scientific-technological fields (including genomics, nutrigenomics, nanotechnology, converging technologies) for national (ministries, advisory bodies, regional development organisations, TA-offices, etc.) and international organisations (European Commission, OECD, UN).

In February 2009, she left TNO and joined the Technopolis Group in Amsterdam where she continued her work as a senior consultant in the field of technology and innovation policy. She has published a number of scientific articles and presented her work at international conferences in the field of technology assessment and innovation policy. She is a member of the editorial board of the International Journal of Biotechnology.

Het onderzoek in dit proefschrift werd financieel ondersteund door de Stichting Agro Keten Kennis (AKK), de Nederlandse Organisatie voor Toegepast Natuurwetenschappelijk Onderzoek (TNO) en het Ministerie van Landbouw, Natuur en Voedselkwaliteit (LNV).

Voor de financiële ondersteuning van de vermenigvuldiging van dit proefschrift gaat erkentelijkheid uit naar de Wageningen Universiteit.

Innovation and sustainability series

The fields of innovation and sustainability are more and more recognized as the major drivers of business success in the 21st century. Today's companies are facing ever-faster changes in their business environment, to which they must respond through continuous innovation. The growing concern regarding the quality and environmental friendliness of products and processes call for fundamentally new ways of developing, producing and marketing of products. New ways of organizing supply chains, with new network ties between firms are needed to cope with these new demands. This series aims to assist industry to conduct the (interorganizational) innovations needed to meet the challenges that are fundamental for the transition from a production orientation to a 'cradle-to-cradle' demand-orientation. However, innovation can be disruptive, not only concerning the organization of the processes, but also regarding the allocation of resources and power bases. Existing companies are increasingly challenged by newcomers, e.g. start-up firms and spin-off ventures. In the transition process, supplier bases might be reorganized, activities reallocated, and relations and role allocations changed as new entities occur. We want to study these new organizational forms and their consequences – as we view them as core for these business networks in transition.

About the editor

Onno Omta is chaired professor in Business Administration at Wageningen University and Research Centre, the Netherlands. He received an MSc in Biochemistry and a PhD in innovation management, both from the University of Groningen. He is the Editor-in-Chief of The Journal on Chain and Network Science, and he has published numerous articles in leading scientific journals in the field of chains and networks and innovation. He has worked as a consultant and researcher for a large variety of (multinational) technology-based prospector companies within the agri-food industry (e.g. Unilever, VION, Bonduelle, Campina, Friesland Foods, FloraHolland) and in other industries (e.g. SKF, Airbus, Erickson, Exxon, Hilti and Philips).

Guest editor

Felix Janszen is part-time professor Management of Technology and Innovation at the Erasmus University Rotterdam. He belongs to the breed of entrepreneurial academics and has worked always at the interface of academia and business. He is co-founder of 2 biotech companies and the Center of Innovation Management and founder and CTO of Inpaqt bv, a spin-off of the Erasmus University. Inpaqt helps organizations to increase their innovation capabilities and speed up their innovation process using a unique software toolbox 'the Innovation Management Suite'. He has studied biochemistry (University of Leiden and the Erasmus University of Rotterdam) and business economics (University of Amsterdam). His main interests are in complex, non-linear systems and developing tools that helps to facilitate the non-linear dynamics in real life situations.

Printed in the United States
by Baker & Taylor Publisher Services